IMPROVING TV SIGNAL RECEPTION

MASTERING ANTENNAS AND SATELLITE DISHES

DICK GLASS

TAB BOOKS
Blue Ridge Summit, PA

FIRST EDITION
FOURTH PRINTING

© 1988 by **TAB BOOKS.**
TAB BOOKS is a division of McGraw-Hill, Inc.

Library of Congress Cataloging-in-Publication Data

Glass, Dick.
  Improving TV signal reception : mastering antennas and satellite
  dishes / by Dick Glass.
      p.   cm.
  Includes index.
  ISBN 0-8306-0270-4   ISBN 0-8306-2970-X (pbk.)
  1. Television—Antennas. 2. Earth stations (Satellite
  telecommunication)   I. Title.
  TK6655.A6G58   1988   88-5468
  621.388'5—dc 19   CIP

TAB BOOKS offers software for sale. For information and a catalog, please contact
TAB Software Department, Blue Ridge Summit, PA, 17294-0850.

Questions regarding the content of this book should be addressed to:

Reader Inquiry Branch
TAB BOOKS
Blue Ridge Summit, PA 17294-0850

Contents

Introduction

Why would anyone want to become involved in the antenna, satellite or MATV business? After all, few schools teach the technical aspects of it. Wouldn't you rather be a: computer, TV, office machine repair, VCR, communications, radio/TV store operator, process control, or CNC technician.

These are just a few of the jobs at which technicians work. Schools have done a wonderful job of producing electronics technicians, but now it is not so easy to find comfortable positions in computers, CNC, or process control. Mass merchandisers have crushed the urban radio/TV store technician/operator. VCR and TV prices are so low that new sets are available for less money than the bill for many repair jobs. Office machine technicians aren't getting rich, and the jobs frequently offer little opportunity to get experience. Communications radio is currently a better place for technicians and shopowners, but the product is becoming more reliable and less in need of circuit troubleshooting than in the past.

Each of the above jobs requires much knowledge and experience. The financial rewards are no longer what they once were. There is a lot of work to be done by technicians, but is it as exciting as it used to be? Are the financial offerings as enticing? No. In many cases they are just jobs.

Satellites, antennas, and MATV systems require technical knowledge and experience, and the electronics theory isn't difficult. It is much more interesting than fixing TVs and VCRs. The public has a growing desire for more TV programming. Millions of homes need what the SAM technician knows. The field is wide open. In most areas it is now difficult to find a competent antenna technician, pre-wirer, tower technician, satellite repair technician, or master antenna specialist. People are willing to pay you to help them get more TV signals.

Can you learn SAM work? Yes.

Can you make more money in SAM work? Yes.

Will it take months and months of study? No.

This book can give you the basic ideas. With a little effort and a thirst to learn all about the SAM business, you can find a most interesting and rewarding career bringing a much-needed service to a public that is willing to pay you.

Good luck.

Chapter 1

The SAM Business:
Satellites, Antennas and MATV

Can you think of a business or technology that has changed as fast or to the extent that radio and TV have? Has housing changed a lot in a hundred years? Or auto making; accounting; trucking; food service or farming?

A SHORT HISTORY OF RADIO AND TV

In the 1930s and 1940s the well-to-do American family proudly showed off its console radio as the centerpiece of the front room furnishings. How exciting it was to listen to the world news, a world championship boxing match, or a national political campaign. While some of us got serious about short wave radio, the general public was having a romance with AM Radio, 550 to 1630 kilocycles (we called it kilocycles rather than kilohertz in those days). The AM radio and the telephone became household necessities. Nearly every home had one or more. AM auto radios followed, becoming as important an accessory as an electric starter or windshield wipers.

With the introduction of high-fidelity sound and stereo, broadcasters improved the quality of broadcast sound. What a thrill the first time you experienced a stereo presentation. Radio stations quickly began stereo FM service and a whole new frequency band came into use by the public. The fast-growing audience, and the public's thirst for more and better radio programming caused advertisers to chase the "golden carrot," which generated more stations and more diversified programming. It reached an ever larger portion of the public. Broadcasting became big business.

It wasn't difficult to recognize that TV broadcasting could surpass radio in attracting viewers and listeners. TV became even bigger business. It was a quantum leap in the

1

ability to communicate information. Everyone knew it. Even the early black and white programming was a phenomenon that seemed to mesmerize people. The networks brought world-wide news and events; advertisers and programmers produced entertainment and cultural products which quickly introduced us all to new ideas and new ways. Just as our ears have become "educated" and our standards for sound have risen ever upwards, our eyes and minds have become "educated" to demand higher TV programming standards in the past 30 years.

It seemed forever before the black and white image on TV was replaced with color. You can probably still remember the first time you saw a color TV picture. It was so much better you just had to have a color TV. Now everyone has one, or two, or three.

When TV first became available in the late 1940s and early 1950s, you may have watched one local station. If you were near a big city, you probably used a rabbit-ear antenna to view that one station. After a while, the one station became boring. Competition came along in broadcasting and you had two or more stations, probably vhf. Perhaps a city close by got a new TV station, and you could receive it, in addition to your local channel. You could watch only one station at a time, but wasn't it nice to have a choice?

Many people that lived close to the broadcast stations, kept using rabbit-ears, because they could get an acceptable picture with them. Most didn't invest in a rooftop antenna at first—after all they were expensive. (Whether the cost of an antenna was $19.95, $199.95, or $499.95, it was still expensive to most of us.) So a great many people just kept on using the "ears." If you finally "gave in" to the idea of TV, and lived out in the country, you had little choice but to get an outside antenna. The best reception often required a tower and rotor. If you were 40 miles from the station you might have had to settle for a pretty snowy picture. A majority of rural households did invest in an antenna, and TV became the major form of family entertainment.

With such success, more stations were added. More uhf channels were allotted and more areas now had both uhf and vhf stations. Many households had either a vhf antenna, or a uhf antenna, but not both. Since it is easier and less expensive to do nothing, that's just what the majority of people did. A uhf station in a formerly all vhf city found it difficult to attract an audience. After all, many people reasoned, who needs to put up a $50 antenna when you already have four vhf stations? Besides, what would you want to watch on that new PBS station, anyway?

Now with even better and more diversified programming on an increased number of channels, most of us decided that maybe a wide reception would be good. So a lot of people improved their antennas to include the missing band (uhf or vhf). Many didn't. Many also hung onto their bent rabbit ears with the big wad of aluminum foil on the ends. This wasn't the best reception, but it was cheap, even if it did have a smeary and ghosty picture. Manufacturers of the ever more popular portable TVs included built-in rabbit ears. It seemed good enough for many people.

With the poor quality antennas, and with many rooftop antennas being inadequate, cable systems found a most susceptible market for their better quality picture and usually a larger number of channels. Psychologically, it was easier to pay a small monthly fee than to invest in an outside antenna, even if it cost more in the long run for cable service.

Where cable wasn't available, especially in the country, people found they could improve reception by adding signal boosters. Transistor technology made these practical. The booster brought quality reception to people living 40 to 50 miles away from the broadcast towers. Only about 2 percent of the population of the U.S. is unable to receive one or more TV stations now.

All the while, from the 1940s to the 1980s, the broadcast industry was upgrading the transmission quality as well as the programming. The diversity and the quality of American TV programming is a human cultural phenomenon. Compared with many other areas of the world, the United States and a few other countries offer a fantastic array of broadcast material. The educational and entertainment value to the people is immense.

SATELLITE TV

In the late 1970s, satellite TV became an interesting curiosity. The idea seemed plausible, but the cost for the average person was entirely too high. After all, what could you get up there that wasn't available otherwise? The answer was . . . a lot. All of the sudden, it seemed, instead of the availability of a half-dozen or so channels, over 100 could be received from the satellites. By 1987, over two million households were equipped to view this new form of communications that could reach even the last 2 percent!

TV communication is so accepted now that even recreational vehicles and small boats are equipped with a dish. Think about the popularity of TV . . . some people now carry portable, battery operated, or hand-held TVs!

The habits and attitudes of people have changed. The insatiable appetite for movies, sports events, news, information and weather, even data communications, has forever left those who resisted television (as unnecessary or evil) far behind. Perhaps the video cassette recorder and disc players helped solidify the attitude that more sources of programming are desirable. With all the forms of TV and radio programming, few multiple dwellings remained for long without adequate signal distribution systems. MATV and SMATV systems now have become a necessity, not just a nicety. Motels and apartment buildings were among the first to add satellite TV for their residents.

Practically unnoticed in the early days of TVRO was the ability to receive radio station subcarriers on many of the satellite channels. Few satellite owners seemed aware that several dozen FM radio stations could be received (from Chicago, Salt Lake City, Los Angeles, New York, etc.) by anyone with a dish. To receive a Salt Lake City radio station in Indianapolis would have been a communications marvel in 1935. Today it is accepted merely as a curious, if seldom used, nicety. Few people realize that practically all radio network broadcasts now deliver their signals by satellite to local radio stations. The communications explosion has left these radio stations less important to some people.

THE ATTITUDE CHANGE

What all this has led up to is that the present-day attitudes of people can be described as a thirst for higher and higher levels of video entertainment. No longer is the public satisfied with one or two snowy channels, used just to catch the evening news. The majority want studio quality reception of all available off-air channels, as well as satellite signals, either direct or from the cable.

HOW THIS AFFECTS THE TECHNICIAN

Antenna technicians, MATV, satellite installers, and the remainder of the electronic service profession, would do well to recognize the trend. It is much easier to sell MATV systems now, even for single family dwellings, than it was only 10 years ago. Everyone wants better reception, and they want all vhf and uhf channels that are within range. People want satellite TV, and most are willing to pay for quality equipment. Even city dwellers want TVRO. Commercial establishments that previously were content to have a TV above the bar that could show an occasional network sports event now want satellite, or at least cable, and nothing less. A new standard has arisen, and it is a high standard.

WHO CAN DO THE WORK?

The only people who can satisfy the need are technicians, capable of understanding, installing and servicing SAM equipment. Where the work was often assigned to "part-time firemen" in the 1950s and 1960s, it is now done by professionals using modern hardware and test equipment. They survey the site, and calculate the exact needs prior to selling or installing any type of system. These professionals work with simple rooftop antennas as well as deep-fringe models. They learn to cancel out unwanted signals, to deal with terrestrial interference, to program exotic equipment and to hook-up apartments with up to hundreds of taps. The new SAM professional is a reception specialist whose qualifications include expertise as a programmer of various brands of computerized equipment, troubleshooter, instrumentation technician, cable splicer, mechanic, and an electrician with an understanding of building and home construction. He must be knowledgeable in several areas of communications, and completely familiar with home entertainment products and programming.

That's what this book is about. Helping you to overcome any fears you have about your ability to service, troubleshoot, and install SAM equipment. Your services are in demand. You are doing a job that takes an expert to do right. Because there are presently very few technicians proficient in all these areas, you should find being in business for yourself, or working for a SAM dealer, very rewarding.

QUIZ

1. In the early days of radio, and the days of cylinder phonograph records, sound reproduction at frequencies between 200 cps and 4000 cps was accepted as enjoyable entertainment.

   a. ( ) true
   b. ( ) false

2. Within a few years, people started buying radios and phonographs which would extend the audio reproduction range to include more bass sounds and high notes.

   a. ( ) true
   b. ( ) false

3. Within a few years, people started to recognize that booming bass was unnatural. High Fidelity sound not only produced a range of audio that is near the human hearing range, but also came close to eliminating unnatural mechanical resonant

distortion. It also improved the balance between highs and low levels and high and low frequencies.

a. ( ) true
b. ( ) false

4. Within a few more years, people started recognizing that high fidelity sound was missing a certain depth or presence that live music has. Stereo sound provides depth, and a perception of direction that makes recordings and radio broadcasting even more natural.

a. ( ) true
b. ( ) false

5. So far as the human anatomy is concerned, something of an educational nature took place over a few decades regarding human sound perception. The evolution of the educational process mostly involved _____

a. ( ) hearing
b. ( ) the mind

6. In a similar manner to the hearing perception change, people recognized that black and white video with snowy and ghosty pictures as inferior because these reproductions . . .

a. ( ) caused eye pain
b. ( ) were not true or natural

7. After becoming accustomed to perfect moving pictures, high quality cable TV and VCR programs, most people were content to continue watching snowy, and ghosty pictures.

a. ( ) true
b. ( ) false

8. Fortunately for consumer electronics dealers who are involved in reception products, antenna technicians can be trained easily to do the work of installing and servicing antennas and MATV systems.

a. ( ) true
b. ( ) false

9. Until recently, which of the following provided training courses preparatory to becoming a SAM technician?

a. ( ) commercial electronics associate degree schools
b. ( ) vocational high schools

c. ( ) engineering colleges
d. ( ) reception equipment manufacturers
e. ( ) none of the above

10. With the public attitude demanding increasing diversity of programming, and improved quality of reproduction of TV and audio communications, the total dollars generated in the SAM business, the man-hours needed to install, service, and improve reception equipment will . . .

a. ( ) increase
b. ( ) decrease

Chapter 2

Broadcast Frequency Bands

Communications technicians ordinarily work with only one or two bands of frequencies. In their daily routines they learn the peculiarities of these frequencies. It is advantageous to become familiar with the other frequency bands used for various other communications services too. The "complete" technician is aware of the major bands in the rf spectrum as assigned by the FCC and international regulatory bodies. The technician should know what communications and what type of programming can be expected in each band and the possible effect one type of transmission can have on another.

RF COMMUNICATIONS FREQUENCIES

When we speak of frequency bands, we are describing the frequencies used in radio, TV, and other forms of electronic communications. These frequencies start at one cycle per second, then progress through the audio frequencies of 20 to 20,000 cps (or Hertz), then into the nine radio bands, all the way into the range of visible light. Above 1 gigahertz is generally considered to be *microwave*.

The frequency of visible light is 3.8×10^{14} to 8×10^{14} Hz. This is 380,000 GHz to 800,000 GHz. Infrared frequencies begin at 300 GHz and extend to the lower end of the visible range.

Radar technicians use one or more of the radar bands, TV technicians use the three TV bands, and the cable bands, Satellite technicians use the C, or Ku bands of frequencies.

Because the assignments of the frequency spectrum by the FCC began in the early days of radio and no one imagined we would eventually use practically all of it, the assignments are somewhat disorderly. Some services can operate in a narrow band of frequencies, while others, like TV and satellite transmissions, need large bandwidths to facilitate transmission of the complex signals required. As an example, the AM radio broadcast band is only about 1.07 MHz wide (535 kHz to 1605 kHz). As you tune your AM radio through the AM band, it seems this major entertainment and news service band must surely be taking up a large portion of the available broadcasting frequency range. In truth, the total AM radio band, being about 1 MHz wide, is only one-sixth as large as a single TV channel (which is 6 MHz wide). There are 82 TV channels plus many more cable channels!

In Table 2-1, we show the basic classification of the spectrum. By designating the bands in logical increments and attaching a name to them, we can understand the band location for the equipment we are working on or servicing. Vhf, for instance, is going to be between 30 MHz and 300 MHz. Therefore, channel 2 TV stations are vhf, operating at 54 to 60 MHz. The FM broadcast band of 88-108 MHz is also in the vhf band (Table 2-2).

Before we delve into a familiarization study of the radio frequency spectrum, we technicians should have at least a cursory understanding of the *very low frequency* band, which is the audio spectrum. Note that humans can generally hear from about 20 Hz to 16,000 Hz. AM radio, all these years has had audio reproduction covering about 50 Hz to 8 kHz. Since the radio broadcast signal is composed of a carrier, modulated by a continuous stream of audio, the highest pitch, or frequency of audio, modulating the transmitter carrier frequency will cause that transmitter carrier frequency to deviate from the exact carrier only 8 kHz on either side, or 16 kHz total. The bandwidth of the AM station then is 16 kHz. In TV, the complex video signal isn't a continuous stream of video, instead it is video, interrupted at the end of each of 525 lines, by sync (locking) pulses. At the end of each of 60 frames it is also interrupted by vertical sync pulses, color stabilization information and test signals. Without any actual video information (just a white screen), the TV signal requires over 31 kHz. If you add 100 dots of video information on each line you need 3.1 MHz. Add in the FM sound carrier and sideband of + or − 25 kHz, and you see why a TV channel takes up so much spectrum space.

"Ham" radio bands are another method of attempting to put segments of the spectrum in terms we can visualize mentally, and relate to. Table 2-3 lists the common amateur radio bands.

Table 2-1. Frequency Band Classifications.

Frequency		Classification	Abbreviation
3-30 kc		Very low frequencies	VLF
30-300 kc		Low frequencies	LF
300-3000 kc		Medium frequencies	MF
3-30 mc		High frequencies	HF
30-300 mc		Very high frequencies	VHF
300-3000 mc		Ultrahigh frequencies	UHF
3000-30,000 mc		Super-high frequencies	SHF
30,000-300,000 mc		Extremely high frequencies	EHF
300,000-3,000,000 mc			

Table 2-2. Audio Spectrum of Frequencies.

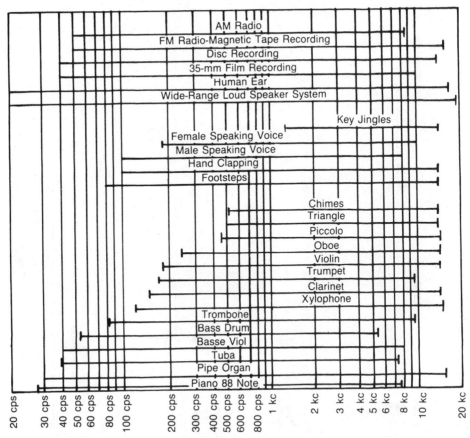

We have mentioned the audio spectrum, the "ham" bands, the AM and FM radio bands, TV and satellite, plus infrared and visible light. One might assume this then is the bulk of the spectrum in use. These bands are very important, but there are many more segments of the spectrum assigned to special purposes. For instance, ocean-river navigation, aircraft radio, weather and scientific radio are all important uses. The government operates WWV and WWVH time standards on several different frequencies (as do other countries). Most of these time signals are in the High Frequency band.

Table 2-3. Frequency Bands Using Meter Identification.

Band	Frequency (mc)
80 Meters	3.5-4.0
40 Meters	7.0-7.3
20 Meters	14.0-14.35
15 Meters	21.0-21.45
10 Meters	28.0-29.7
6 Meters	50-54
2 Meters	144-148

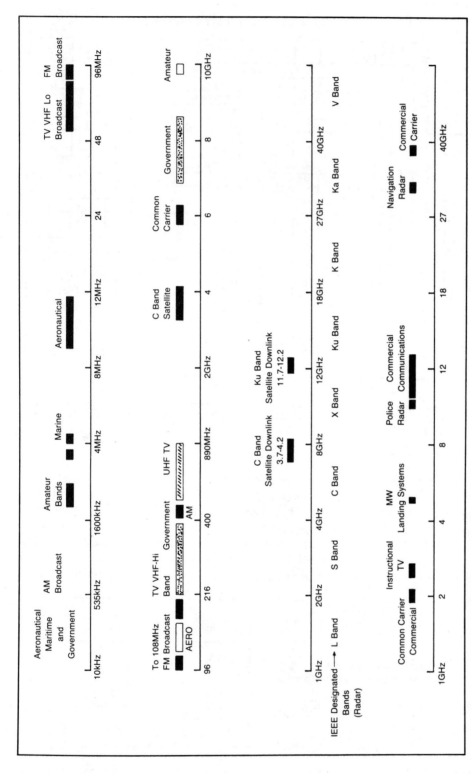

Fig. 2-1. Chart of 10 kHz to 40 GHz frequency band allocations.

The U. S. Government has set aside for its own use for commerce, aviation and the military over two dozen separate bands of frequencies. Telegraphy and telephony are important usages and have several assigned bands. Radar is used by aircraft, ships, airports and harbor facilities as well as industrial firms, mostly utilizing uhf and shf band segments.

Telemetry is now commonly used for data transmission. Unique spectrum spaces have been assigned to both land-based and satellite transmission for this purpose.

To attempt to memorize all of the services and frequencies used would be a difficult task. The most practical answer then is to know the major bands, to identify their ranges (i.e. vhf = 30 to 300 MHz), and to get a mental picture of the location of the bands you are likely to be working with, or that affect you. You should know the audio range, the AM and FM radio bands, the low and high ranges used by TV, uhf and perhaps the C-Band satellite downlink band. Once you know these, any other band you need to know can be remembered, as they relate to these common bands. Obviously, if your job has to do with many different services, you need an official and complete chart showing the FCC assignments for all types of transmissions.

The FCC has set out rules for the types of transmissions that can be used, the frequency limits for any one station, permissible power and antenna height, distance between co-channel transmitters, hours of daily service, and other limits. You should have a cursory knowledge of the rules governing use of the rf spectrum in general, and a complete knowledge of rules governing the bands you are involved in.

One final reason it is difficult to know the locations of all the services is that some bands are used for infrequent communication, or for several types simultaneously. Examples are the fixed, mobile, aeronautical, or maritime services. Some services are reserved strictly for one purpose. For instance, 156.7625 MHz to 156.8375 MHz is reserved for maritime distress only.

Figure 2-1 is an extensive list of all the bands and services presently in use. By keeping this book, or the chart handy, you can have a ready reference for all the rf frequency bands.

QUIZ

1. The human eye and the human ear utilize approximately the same bandwidth of frequencies to interpret aural or visual information.

   a. ( ) true
   b. ( ) false

2. The broadcast AM radio band takes up approximately 50 percent of the amount of rf spectrum space as that utilized by the FM radio band.

   a. ( ) true
   b. ( ) false

3. The spectrum of frequencies that produce infrared light is below the frequencies of visible light.

   a. ( ) true
   b. ( ) false

4. The U. S. Government standard time signals are broadcast on WWV and WWVH in the high frequency band.

   a. ( ) true
   b. ( ) false

5. The low frequency band ranges from 30 kHz to 300 kHz.

   a. ( ) true
   b. ( ) false

6. The medium frequency band ranges from 400 kHz to 3 MHz.

   a. ( ) true
   b. ( ) false

7. If your transmitter broadcasts a carrier frequency of 100,000 cycles per second and it was never modulated with audio or other information, the bandwidth would be about _____

   a. ( ) 1 Hz
   b. ( ) 8 kHz

8. If your 100,000 Hz transmitter carrier frequency was modulated with a 60 Hz line frequency signal, the bandwidth would be from 99,040 to _____

   a. ( ) 100,060 Hz
   b. ( ) 100,040 Hz

9. The "Ham" bands are all concentrated in the vhf band.

   a. ( ) true
   b. ( ) false

10. Communications technicians must know the frequency allotment and services utilizing all the rf spectrum bands.

   a. ( ) true
   b. ( ) false

Chapter 3

Antenna Resonance

Understanding antenna resonance has not been easy for most technicians. Working the formulas and figuring wavelength for a certain frequency signal is one thing, but getting the "feel" of what is really going on is something difficult.

In this example, we will use as a reference a signal we are all somewhat familiar with, the TV channel 11 frequency. That carrier is at 199 megahertz.

The formula for wavelength is:

$$\text{Wavelength (feet)} = \frac{984}{\text{f in MHz}}$$

Working the formula for channel 11:

$$\text{Wavelength} = \frac{984}{199} = 4.94 \text{ feet}$$

Then ½ wavelength for channel 11 is 30 inches (2.48 feet).

Conceivably there are ways to make a full wavelength antenna work. However, it will be a rare occasion that you will find any antennas used in TV, other than ½ wavelength in size. A ½ wavelength antenna will respond to the electromagnetic and electrostatic waves traveling from the broadcast antenna to it, provided both transmitter and receiving antenna are parallel to each other. A TV antenna twice as long as the channel

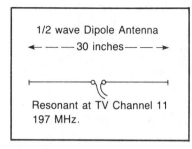

1/2 wave Dipole Antenna

←— — —30 inches— — —→

Resonant at TV Channel 11
197 MHz.

Fig. 3-1. The ½ wave dipole antenna, resonant at TV channel 11, 197 MHz.

11 antenna used here, would respond to a frequency ½ that of channel 11, or about 99.5 MHz. The ½ wavelength antenna used in television is capable of reacting to the influence of the broadcast waves so that it can have a maximum voltage at one point on its length while at the same time having a minimum voltage at another point. Thus the output terminals can produce a voltage that is a small replica of the originally transmitted one.

The basic antenna from which most TV antennas evolved is shown in Fig. 3-1. It is called a ½ wave dipole antenna. In Fig. 3-1 note that the antenna is about 30 inches long which is ½ wavelength at the frequency of channel 11.

In Fig. 3-2 note that the voltage levels are maximum at both ends of the antenna rods when they are minimum at the center feed connections. One quarter-wavelength later (or earlier) the voltage levels are minimum at the ends and maximum at the center. The channel 11 signal waveforms from the broadcast station resonate the electrical particles in the rods. Just like the wind blowing past a length of metal tubing, it may "hum," vibrating mechanically at its *mechanical resonant frequency*. The *electronic resonant frequency* is not exactly the same thing, but it can be visualized in the same manner. As to how the simple rods deliver an actual electrical voltage impulse of several hundred microvolts at the terminals, think of the channel 11 broadcast signal causing an inductive influence on the electrons in the metal of the rods. This causes a difference of potential from one point on the rods to another, from one end to the other.

In Fig. 3-2 notice how a full wavelength is exactly twice the length of the antenna rods and that ½ wavelength acts the same as a full wavelength. We can stop using a full wavelength in discussing this subject now. Note in Fig. 3-3 how the voltage levels

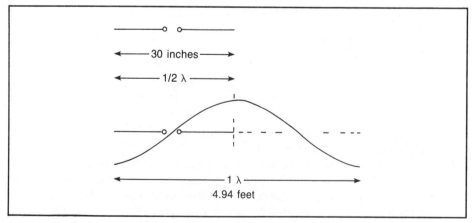

Fig. 3-2. Dipole antenna showing length at ½ wavelength, compared with full waveform.

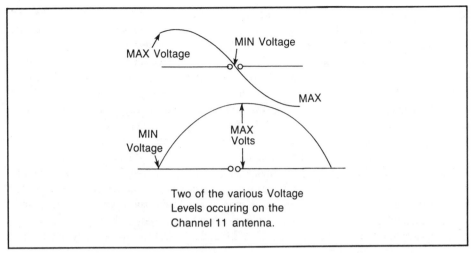

Two of the various Voltage
Levels occuring on the
Channel 11 antenna.

Fig. 3-3. Two of the infinite waveforms occurring on the channel 11 antenna.

change in the time a cycle passes. To try to understand this better, think of the transmitter at channel 11 as the primary of a transformer, and the receiving antenna as the secondary. The channel 11 broadcast antenna causes an inductive influence on the electrons in the metal of the rods—a difference in potential from 1 point in the rods to another—from one end to the other.

Now that you see what happens to the channel 11 antenna when the exact right size signal waveform hits it, let's look at what happens to the same antenna rod when a higher frequency waveform hits it: In Fig. 3-4 you see the same channel 11 antenna

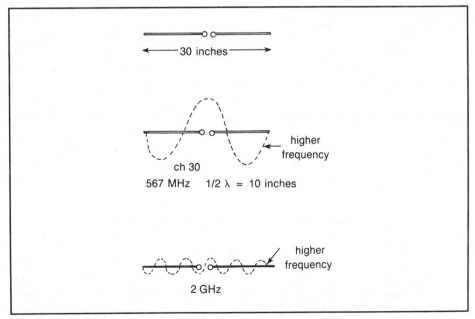

Fig. 3-4. Two higher frequency waveforms appearing on the channel 11 antenna.

rod, but now lets see what happens when a much higher frequency, channel 30, passes it. Channel 30 is at 567 MHz. Using your formula, calculate the ½ wavelength for channel 30: It is about 10 inches. Note in Fig. 3-4 that the higher wavelength doesn't "fit." As the cycles progress, you never have a full wavelength exactly fitting the rods and over 3½ wavelengths can ramble around the rod. The confusion gets worse with a much higher frequency of 2 gigahertz. Now many nodes are rambling across the rods and never is there much difference of potential between one end and the center connections. The antenna is not resonant at the channel 30, or the 2 GHz, frequencies. If any signal is picked up, it will be much lower than the resonant frequency, if it exists at all.

Using the same channel 11 antenna, look what happens with a lower frequency, or longer waveform: Channel 2 has a ½ wavelength of 9 feet. Even ¼ wavelength is too long to fit the 30-inch rods, and you can see there can be practically no difference between the rod ends and the center feed connectors. Again, there will be little, if any, signal produced by the antenna at this non-resonant frequency. Figure 3-5 shows the quarter wavelength of channel 2; while Figs. 3-6 and 3-7 illustrate the analogy of matched antennas to a transformer.

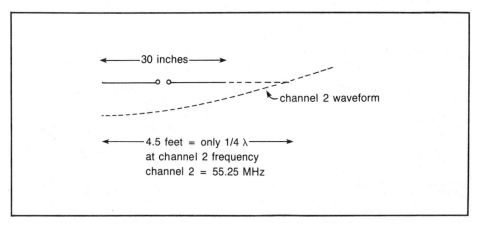

Fig. 3-5. A much lower waveform appearing on the channel 11 antenna.

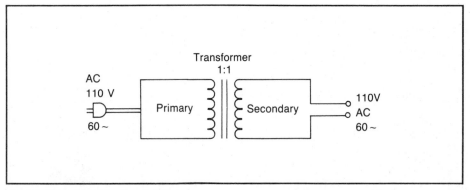

Fig. 3-6. Schematic of a transformer.

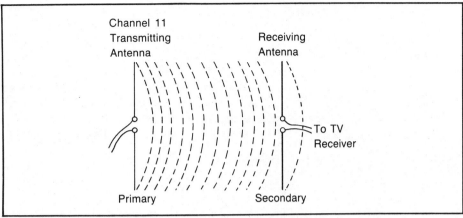

Fig. 3-7. Schematic of a TV antenna comparing it to a transformer.

QUIZ

1. Which musical instrument has the highest mechanical resonant frequency?

    a. ( ) tuba
    b. ( ) piccolo
    c. ( ) bass drum

2. A poor quality loud speaker, 15-inches in diameter, will produce a higher level of output at its mechanical resonant frequency than at other frequencies.

    a. ( ) true
    b. ( ) false

3. A 15-inch loudspeaker can have a resonant frequency near _____ Hz?

    a. ( ) 30
    b. ( ) 300
    c. ( ) 3000

4. The electrostatic and electromagnetic impulses transmitted from the TV broadcast station will cause a maximum magnetic influence in a metal rod positioned _____ to the broadcast antenna.

    a. ( ) parallel
    b. ( ) perpendicular

5. TV antennas are _____ polarized.

    a. ( ) horizontally
    b. ( ) vertically

6. CB antennas are _____ polarized.

    a. ( ) vertically
    b. ( ) horizontally

7. In addition to transmitting radio signal impulses, Nicolae Tesla showed experiments in the early 1900s demonstrating that it is possible to transmit power. While the current that a radio or TV antenna receives is very small, the TV antenna actually is receiving an impulse of both current and voltage.

    a. ( ) true
    b. ( ) false

8. The resonant frequency of a TV antenna at TV channel 13 is _____?

    a. ( ) 211 MHz
    b. ( ) 422 MHz
    c. ( ) 844 MHz
    d. ( ) 1688 MHz

9. The ½ wavelength of an antenna resonant to channel-13 TV frequency is?

    a. ( ) 28 inches
    b. ( ) 31 inches
    c. ( ) 14 inches
    d. ( ) 56 inches

10. A channel-13 TV antenna produced a 100 μV signal at its terminals. An identical channel 2 signal is being transmitted to the channel 13 antenna. What signal might be expected at the terminals?

    a. ( ) 100 μV
    b. ( ) 50 μV
    c. ( ) 200 μV

Chapter 4

Decibels—An Unfriendly Unfamiliar Term

There are few concepts in electronics which don't eventually end up using decibels, especially when you study antennas and signal distribution systems or audio technology. Few technicians are able to handle decibels easily, or quickly understand quantities expressed in dBs. The problem is that decibels aren't common-usage terms (like inches, pounds or gallons) we all can easily visualize. It's hard to visualize 6 dBs. How big? How wide? How many are 6 dBs? On the other hand, decibels really aren't difficult if you spend just a little time getting used to them.

A FEW CONCEPTS CAN CLEAR IT UP

If you get a few basic truths locked into your mind, you will easily be able to figure all the values and levels used in systems, and to understand specifications using dBs.

TWO WAYS

One problem in understanding is we use dBs in two different ways: In TV we establish 0 dB as 1000 microvolts (μV). When discussing signal strength in TV and MATV systems, we can communicate by stating a 100,000 μV signal is also a 40 dB signal. Or, a 500 μV TV signal is a -6 dB signal. So decibels are used in this way to compare a specific voltage level. This is convenient if you are frequently dealing with signals that all relate to one base, or set amount of signal. See Table 4-1.

Table 4-1. Conversion Table—dB to Microvolts.

Reference Level: 0 dB = 1 Millivolt (1000 μV)

dB	μV	dB	μV	dB	μV
−40	10.00	0	1 000	41	112 200
−39	11.22	1	1 122	42	125 900
−38	12.59	2	1 259	43	141 300
−37	14.13	3	1 413	44	158 500
−36	15.85	4	1 585	45	177 800
−35	17.78	5	1 778	46	199 500
−34	19.95	6	1 995	47	223 900
−33	22.39	7	2 239	48	251 200
−32	25.12	8	2 512	49	281 800
−31	28.18	9	2 818	50	316 200
−30	31.62	10	3 162	51	354 800
−29	35.48	11	3 548	52	398 100
−28	39.81	12	3 981	53	446 700
−27	44.67	13	4 467	54	501 200
−26	50.12	14	5 012	55	562 300
−25	56.23	15	5 623	56	631 000
−24	63.10	16	6 310	57	707 900
−23	70.79	17	7 079	58	794 300
−22	79.43	18	7 943	59	891 300
−21	89.13	19	8 913	60	1 000 000
−20	100.0	20	10 000	61	1 122 000
−19	112.2	21	11 220	62	1 259 000
−18	125.9	22	12 590	63	1 413 000
−17	141.3	23	14 130	64	1 585 000
−16	158.5	24	15 850	65	1 778 000
−15	177.8	25	17 780	66	1 995 000
−14	199.5	26	19 950	67	2 239 000
−13	223.9	27	22 390	68	2 512 000
−12	251.2	28	25 120	69	2 818 000
−11	281.8	29	28 180	70	3 162 000
−10	316.2	30	31 620	71	3 548 000
− 9	354.8	31	35 480	72	3 981 000
− 8	398.1	32	39 810	73	4 467 000
− 7	446.7	33	44 670	74	5 012 000
− 6	501.2	34	50 120	75	5 623 000
− 5	562.3	35	56 230	76	6 310 000
− 4	631.0	36	63 100	77	7 079 000
− 3	707.9	37	70 790	78	7 943 000
− 2	794.3	38	79 430	79	8 913 000
− 1	891.3	39	89 130	80	10 000 000
− 0	1000.0	40	100 000		

0 dBmV: TV SETS NEED IT

Why is 1000 µV important? It's because we had to start somewhere. In the early design of TV sets it was determined that a TV receiver should be manufactured so that it would produce a "snow free" video picture with an input signal of 1000 µV, or 0 dB, of broadcast signal. TV set makers couldn't design one set to operate on 50 µV and another on 1 full volt. So 1000 µV or 1 millivolt was the design standard. If you remember that a TV set requires 1000 µV, or 0 dB of signal, you will already know the most important single fact you need to understand: Relative power measurement. When speaking of TV signal levels from an antenna, or in a MATV system, you can refer to any signal level as it relates to 0 dB. Any technician who works on antennas for long will know that a −20 dB signal level is going to produce a very snowy picture and a +20 dB signal will not only be ample to supply a perfect noise-free picture, but could easily be split to supply 2, 3, or 4 more TVs and all would have snow-free picture quality.

Expressing electromagnetic energy as a signal level relative to 0 dB is a custom we have grown to use in TV-radio work. But decibels are used in comparing human hearing, and in audio work, 0 dB sometimes means something else.

THE OTHER WAY

The other way to use decibels in electronics is in expressing the "difference" between one signal level, and another. An easy way of visualizing this is to compare the amount of signal you would expect from two different antennas—one stronger than the other (Fig. 4-1).

One antenna, A, is a basic ½ wave dipole. Other antennas are all based, or compared, with it. The antenna in A, no matter how many microvolts are produced at its terminals, is said to have 0 dB of gain. We had to start somewhere, so we started with a basic rod antenna. It is the standard starting point. Other antennas don't actually have gain in the sense that they amplify the signal received. By stating that an antenna has a 6 dB gain, we mean that it will pick up more signal (twice as much) than a simple dipole. We can't ever say one antenna will have an exact microvolt level of signal at its terminals. That is because the antenna could be sitting directly under a broadcast antenna, thus receiving several volts of signal, or it could be a hundred miles from the antenna and be lucky to receive 50 µV. If we all know that an antenna with 6 dB gain will perform

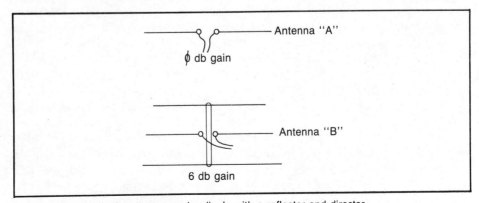

Fig. 4-1. A basic dipole antenna, and a dipole with a reflector and director.

twice as well as a basic ½ wave dipole, we have something to compare with, to visualize, and we can understand it.

A second example would be: you have brought a TV signal into your house. You want to supply a second TV as well as the main set. To divide the signal you need a splitter that properly maintains the impedance of the system (75 ohms when using TV coax). Signal splitters basically all have the same loss, and that is 3.5 dB. Right away you can see that splitters do not divide a signal in half, it's better than that. You only lose about a third of the signal to each output of a 2-way splitter. Knowing that whenever you divide the signal, using a splitter, that you will lose 3.5 dB, means you can accurately determine what signal you will have at the TV sets, or in any other system. It doesn't mean you can state the microvolt level at the TV terminals. It means that you can determine component and wire losses, regardless of what the input signal level is.

In the above example, you want to split the signal to 2 TVs rather than 1. If your antenna was supplying only −10 dB of signal, and you lost another 3.5 dB through the splitter, you would then be left with −13.5 dB at the TV set. Or, if you had +6 dB supplied by the antenna, and lost 3.5 dB, you would still have +2.5 dB at the TV sets. To one familiar with TV signals, the −13.5 dB signal is going to make the original signal (the original, marginal at best, −10 dB signal) now intolerable. On the other hand, the +2.5 dB signal left over from the hypothetical +6 dB signal is still excellent.

It's important to know that these wire losses and splitter losses are something you can count on. You don't need to know how much signal a TV set needs; you don't need to know that the signal level is coming from an antenna, or out of an amplifier, in order to determine a system's losses. No matter what the losses, they remain the same, in decibels, whether you are dealing with a small microvolt signal, or several volts of signal. This makes it quite easy to analyze MATV systems, or to determine what you will have after you install an antenna and run it to three or four rooms in a home. By using decibels to calculate gain or loss in signal systems, the numbers are small, usually only one or two digits. Were we to use microvolt levels we would be confused with millionths of a volt and hundreds of thousands of microvolts. Life is confusing enough! See Table 4-1.

SIX DECIBELS

Six dB is the secret to using dB mathematics. A gain of 6 dB means whatever signal level you originally had has increased to twice as much. A signal level of 1 volt, increased by 6 dB will now be 2 volts. A 200 μV level increased by 6 dB will now be 400 μV. No matter what your original level, a 6 dB level is twice the original minus six decibels result in ½ the original. A 1 volt signal, when reduced by −6 dB is now 0.5 volt (or 500 millivolts). Notice that a +6dB gain was a 1000 millivolt gain, but a −6 dB level is only a 500 millivolt loss. That's important to recognize. The 200 μV level, if reduced by 6 dB is now only 100 μV.

You can calculate signal levels anywhere in a system by simply halving, or doubling voltage levels.

Let's use another example to show how easy it is: Start with a 5-volt signal. After amplification you now have 25 volts. What amount of gain did you produce within the amplifier? Obviously it is 20 volts, however to use decibels to express that gain let's first double 5 volts to 10 volts. That is a 6 dB gain. Now double the 10 volts to 20 volts. That's another 6 dB gain. Now we have 12 dB of gain. Let's double the 20 volts to

40 volts. That would be another 6 dB gain. But wait. We wanted only 25 volts, not 40. Reviewing, 20 volts was 12 dB of gain. 40 volts is 18 dBs. So, 25 volts is somewhere between 12 and 18 dBs of gain. In antenna work, knowing that might be all you need to know to solve a problem or design a system, but I'm sure you can come closer to the actual dB gain. We know it isn't half the way, or distance between 20 and 40 microvolts, which would be 30 μV. So 25 volts would probably not be a gain of 12 + 3, or 15 dB. It would be less than 15 dB of gain. Let's guess that it would be 14 dB. If you guessed 15, that would be close enough for practically all system calculations. While we can scientifically calculate to an exact dB amount, there is little reason to do so. Why? Because another factor is involved in dB loss calculations. Loss in a TV system will be much less for channel 2 frequencies than for channel 83. Wire loss will be over twice as much for channel 83; therefore, to demand that you be precise in loss calculations accomplishes little. The fact that you don't have to be exact is not a reason for you to remain unfamiliar in using decibels. Think of it as making your job easier and less demanding.

A CHART BASED ON dBmV

Look at Table 4-1. Start by assuring yourself that 0 dB is listed as 1000 μV. It is. Therefore you know all the dB levels shown will pertain to 0 dB, the basic starting point for rf signals. That makes it easy. Now you and I could discuss a signal problem, or we could design a signal distribution system noting the signal levels at any point. We would both know what each was saying to the other.

Analyzing the chart further, note that it only goes negative to −40 dB. Actually that 10 μV signal is not negative. It is truly a +10 μV signal. It is only negative when compared with our standard starting point, 0 dB or 1000 μV.

Why stop at −40 dBmV? I would if I were making up the chart. A 10 μV signal is so small, it is virtually no signal. In fact, the FCC rules governing CATV systems require that any stray signal leakage on the cable system be measurable at no more than −40 dB, that is just about zero signal so far as a TV antenna or TV set is concerned. You've seen far off TV stations appear faintly on a normally unused channel, a signal, barely discernible, usually unable to lock, with so much snow that you cannot recognize objects in the picture, color is nonexistent, and the sound hissing, that is a −40 dB signal! Listing anything below that serves no purpose (Fig. 4-2).

On the other end of the scale the chart shows 80 dB, or 10 volts of signal. That is such a giant signal that most TV sets would be swamped. Preamps and line amplifiers ordinarily cannot handle such a large signal input. Ordinarily, in your work, you will find 35 to 40 dB of signal to be too much. That input is too much for most systems, and too large a signal for a TV set. TVs work great on −6 dB to +25 dB. Anything less is snowy; anything more, and the dreaded adjacent channel interference becomes a problem. Eighty dB though is frequently used in large CATV and cable system trunk lines.

IT ISN'T PERFECT

To illustrate that the use of decibels does not ordinarily require exactness in calculations, in Table 4-1 compare 1 dB with 2 dB. The table shows a 200 μV difference in the levels. Now note the difference between 2 dB and 3 dB. The chart shows only

Fig. 4-2. Indiscernible −40 dB signal as it appears on a TV screen.

a 100 μV difference. Actually 2 dB is a little less than 1300 μV and 3 dB is a little more than 1400. But it's "close enough" for all the work we do. Besides, the chart maker knows you and I can handle even amounts better. That makes comparing easier.

PICKING THE SPECKS OUT OF THE PEPPER

If you really "fall in love" with decibels, understand how useful they can be, and you have a need to be exact, here is a formula for dB calculation when dealing with voltage levels (not power levels).

$$dB = 20 \log \frac{E2}{E1}$$

So we have a voltage level of 200 μV. It is amplified to 1300 μV. Let's do it the easy way first:

$$200—doubled = 400 \ \mu V \ or \ 6 \ dB$$
$$400—doubled = 800 \ \mu V \ or \ 12 \ dB$$
$$800—doubled = 1600 \ \mu V \ or \ 18 \ dB$$

So 1300 is about 60 percent of the difference between 800 and 1600. I'll guess 1300 μV represents 16 dB gain over 200 μV.

Now let's use logs:

$$dB = 20 \log \frac{E2}{E1}$$

$$dB = 20 \log \frac{1300}{200}$$

$$dB = 20 \log 6.5$$

$$dB = 20 \times 0.813$$

$$dB = 16.3$$

Obviously you have to use a logarithm chart to calculate the gain using the log formula. You rarely carry one in your pocket. A lot of us would get confused with the algebra anyway. So I goofed in my dead-reckoning when I guessed 15 dB might be the gain, rather than 16.3 dB. But my guess was close enough for all the antenna work I've ever done!

WATTS CONFUSING?

Now that you have all the above down firmly in your mind, and you really feel that you can handle decibels let's change the rules! When comparing power ratios, the rule changes. Now the dB formula is:

$$dB = 10 \log \frac{P2}{P1}$$

It is not:

$$dB = 20 \log \frac{E2}{E1}$$

This means a gain of 6 dBmW is not a doubling of the power, as it was with decibels when dealing strictly with voltage. A 6 dBmW increase is a quadrupling of the original power. A 3 dB increase is a doubling of the power.

Perhaps you are confused. Why change the rules of the game, right when you thought you "had" it? Well, really the rules haven't changed. The problem is Ohms Law! A watt is made up of both volts and current. Here is the power formula:

$$P = EI, \text{ or Power} = \text{Volts times Current}$$

An example would be: $P = EI$

$$P = 2 \text{ volts times 2 amps} = 4 \text{ watts}$$

Now let's double the voltage and current:

$$P = 4 \text{ volts times } 4 \text{ amps } = 16 \text{ watts}$$

You can see that doubling both the voltage and current would not double the power. Instead it multiplies it by 4.

In your work with power levels, or audio levels, or your own hearing level, a power increase of $+3$ dB is twice the original level. A -3 dBmW signal is ½ the original.

The clue will be the designations used after the dB notation, such as dBmV, or dBmW. You will also see designations that help you more, like dBV, or dBkW (for kilowatts). Most of the time you will be using dBmV.

QUIZ

1. Decibel notation is based on relative values, such as decibels referred to as one millivolt.

   a. ( ) true
   b. ( ) false

2. Basing the dB notation on 1 mV, then 2 mV would be equal to 0 dBmV.

   a. ( ) true
   b. ( ) false

3. If 10,000 microvolts is equal to 20 dBmV, then how many dBmV is 20,000 microvolts?

   a. ( ) 10,000 dB
   b. ( ) 20,000 dB
   c. ( ) 2 dB
   d. ( ) 26 dB

4. When referring to the output levels of an amplifier versus the input level, or an antenna's forward gain versus its rearward signal pickup, decibels are used to indicate:

   a. ( ) current gain
   b. ( ) the ratio between two quantities
   c. ( ) voltage drops
   d. ( ) power supply input

5. Ordinarily, in master antenna systems, it is easier to calculate the line, splitter, and insertion losses by adding, subtracting, multiplying and dividing the microvolt levels, rather than adding or subtracting one or two-digit dB expressions.

   a. ( ) true
   b. ( ) false

Chapter 5

Transmission Lines

The above transmission lines are the types you will encounter in most antenna, MATV, and satellite work. On occasion you will come across other types designed for specific purposes. For instance, Citizens Band radio uses a 50 ohm impedance, RG-58 wire as does other communications equipment. While it may look the same as RG-59 it will not work well. Technicians therefore need to be sure of the transmission lines they are using. Figure 5-1 illustrates the various kinds of cable used in SAM work.

In the beginning, TV sets used "open wire ladder" transmission line. It was simply two copper wires separated every foot by plastic insulators. It was expensive and most difficult to work with. It had low capacitive, inductive, and resistive losses. Properly installed, it worked well.

Open wire was quickly replaced with "twin-lead," 300-ohm ribbon cable. This was much easier to make, ship, handle, and install. Technicians were careful to use insulated "standoffs" on the antenna mast, and to keep the wire away from metal objects.

Wire manufacturers produced other varieties of twin-lead in an attempt to reduce the much greater loss this wire experiences when wet. Perhaps the manufacturers also had in mind that consumers would continue to install wire using improper techniques. Foam was used to encase the two wires, and vinyl was used as an outer covering. This design held the wires away from metal and reduced moisture losses. The drawback to foam and other attempts at optimizing twin-lead wire was that when the cable was made larger in diameter, it became bulkier and harder to work with. It was not flexible, and the two wires frequently broke due to cable rigidity. Foam-insulated twin-lead with shielding was also used for a while. It is still available but is used less and less often.

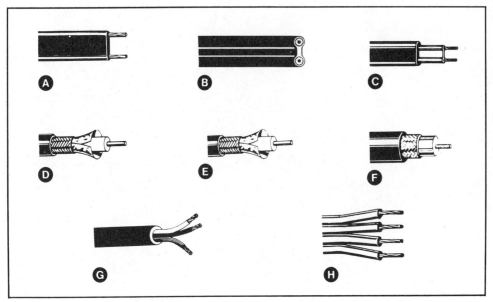

Fig. 5-1. 8 types of wire used in SAM work. a—twin lead wire, b—foam twin lead, c—vinyl clad twin lead, d—RG-59 Coax, e—RG-6 Coax, f—RG-314 or RG-11, g—3-wire rotor cable, h—4-wire rotor cable.

Today RG-59 is the predominant TV cable because mass merchandisers still sell "kit" antennas, which includes twin-lead, it will be around for a while. Technicians and antenna installers must be able to work with this kind of cable for the foreseeable future. Figures 5-2 and 5-3 show how twin-lead and coax cables develop losses.

Theoretically, twin-lead transmission wire has lower losses per 100 feet than RG-59 coax. Standard twin-lead, with #22 conductors, has less than 4 dB loss per 100 feet at 500 MHz, while RG-59, with a #22 center conductor, has a 5.5 dB loss.

Generally, the larger the conductor size, the less loss. It is a good idea to read the specifications on the wire you use, and to stick with a known type, because all coaxial cables are not the same. The early RG-59U, solid coax, using copper braided shield, had a 14 dB loss per 100 feet at the top end of the uhf band (900 MHz). That is nearly twice the loss of foam coax. Braided shield also tends to have a slight radiation ability, which is detrimental. Aluminum foil shielding causes essentially zero radiation loss.

While twin-lead, theoretically, has lower loss than coax (about 40 percent less at 900 MHz), it has so many more problems, its use should be carefully considered. With coax you can depend on the loss specifications. As an example, measure the signal at the antenna: Let's use a +6 dB on our weakest desired signal. We need 40 feet of coax from the antenna to the house entry point, and 60 feet to the farthest TV outlet. Using RG-59 foam, we know there will be 0 dB at channel 19 (500 MHz) and 2 dB at channel 83. With twin-lead, strange things seem to happen. You will find set owners who tape the lead to the mast, dangle it on the ground, lay it in water puddles, pinch it under aluminum windows, run it down walls which have aluminum-backed insulation panels on one side, or frequently run it parallel to power wires. They invariably pull it out beside metal electrical wall outlet boxes. Twisting the twin-lead would help reduce line pickup, but set owners rarely do that. In attics, crawl spaces, and basements the crimes continue

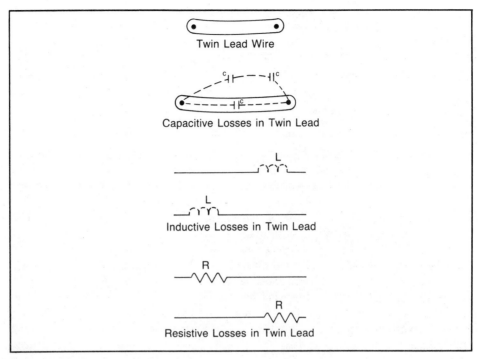

Fig. 5-2. L, C, R losses in twin lead wire.

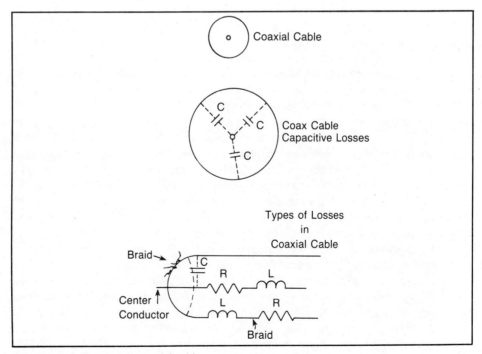

Fig. 5-3. L, C, R losses in coaxial cable.

Table 5-1. Wire Loss Chart.

Wire type	Frequency	Attenuation/100 ft	
	100 MHz	500 MHz	900 MHz
Twin lead	1.4 dry 8.0 wet	4 dry 20 wet	6 dry 30 wet
RG-59	3.0	6.0	8.0
RG-6	2.0	5.0	7.0

as the twin-lead is held out of the way with metal staples, frequently cutting one or both conductors. The twin-lead may be attached to a masonry wall, stuffed behind a furnace, coiled up and laid on a heat duct, wound around a gas or water pipe, or incorrectly spliced or split.

In older installations the twin-lead signal splitters were not designed for uhf. The loss at these frequencies was nearly total!

Whether twin-lead or coax is used, you will frequently find the wrong value of wall taps. While the wall plates and the outlets look neat, the result is often not satisfactory. As a technician you will have to troubleshoot these. The simplest wall outlet is a direct connection, either two screws for twin lead, or a "barrel" connector. These should be terminated with a 300 ohm resistor for twin-lead systems or 75 ohm terminator resistor for coax. Without termination, especially in urban areas with strong signals, standing waves and other problems are possible. A TV set can be used as a terminator. Prewired homes and buildings frequently have the wrong wall outlet taps installed. The most common value and one that often will work in urban areas, is a −17 dB tap. Were you to use that tap in the example above, you would end up with a −17 dB signal. That provides a class B picture, very snowy. A quick solution to that problem would be to substitute a 0 dB outlet at that point.

The other problems with twin-lead are that splitters, taps, and amplifiers are not universally designed for twin-lead applications. There have been attempts to make a full line of these parts, but they are rarely available, and frequent problems will be encountered with them. It is more cumbersome to add in-line amplifiers to twin-lead systems. On the other hand coax fittings are universally available and have known specifications. Today twin-lead is sold mainly through mass merchandisers and is not made to close specifications. Technicians have found that homes pre-wired with modern twin-lead exhibit lower quality signals than expected. These leads are more sensitive to interference and electrical noise. Some installations will be found to have several runs of twin-lead simply spliced by twisting them together. This destroys the impedance match, and sets up ghostlike ringing, or standing-wave conditions. Technicians try to replace twin-lead systems with coax; however, this is not easy where the original twin lead is stapled in the walls, or pinched with insulation, wiring, or wood braces. It may be best to disconnect the twin-lead, and run new coax along different routes.

RG-6 coax has less loss per 100 feet than RG-59 (Fig. 5-1-e). It is commonly used for MATV system work where long cable runs are necessary. Generally speaking, RG-6 is not much higher in cost than RG-59. It can usually be acquired for about 10 percent more. Its use is economical, and the extra dB or so per 100 feet may be all that is needed to provide the proper signal. Some systems use even lower loss coax, RG-11 and RG-214. The RG-214 has only 5 dB of loss at 900 MHz.

RG-11 and RG-214 cost much more than RG-59 or RG-6 (about 15 cents per foot for RG-214 compared with 6 cents for RG-6). These types are usually used only in cable systems, or large apartment complexes. Because of the larger bulk, special end fittings are necessary. These are more difficult to install than "F" connectors, as used on RG-6 and RG-59. Splitters and taps are less universally available also.

The RG-6 has a larger diameter than RG-59. The center conductor is a larger gauge. The center wire for RG-6 is #18, compared with #20 or #22 for RG-59. Many female receptacles will not accept the #18 center conductor. This can cause the technician to attempt to force it into the center hole. Sometimes this works, especially if the center wire is snipped off at an angle, forming a sharper point. If this technique doesn't work, the solution is to screw the F6 fitting onto a standard barrel splice that has a spring-like slit inside the center hole, and easily accepts the #18 wire. The barrel gender is then wrong, so it must be screwed onto a male-to-male F-59 connector, then it can be connected to the smaller female fittings.

The RG-6 also requires a larger-diameter-inner connector. These are common and available, but many beginners in antenna work, and many do it yourselfers will try to install the F-59 connector on RG-6. This is possible, but the foam center insulation must be shaved down and the "F" connector forced on carefully. This practice is not recommended. The possibility of a shorted or loose connection is great. Fortunately the F-6 and F-59 connectors have the same threads, so both fit all the MATV and satellite hardware.

For satellite systems, RG-59 can be used for downconverter-to-receiver cable run. The single conversion receivers operate at a low 70 MHz frequency. For block systems, that return either 450 to 950 MHz, or 950 to 1450 MHz, RG-6 is necessary. Using RG-59 instead of RG-6 (as one might when upgrading to a block satellite system), you might experience sparklies and snow, especially on the lower channels. All the rules for wire are not necessarily true for satellite systems.

Some satellite systems have been designed taking into account the wire losses, so don't blame wire loss for all satellite system signal problems. Frequently, on short runs of wire with large reflector dishes and high gain LNAs, the solution to a snowy picture is to install a 12 dB attentuator in the signal coax line from the dish downconverter.

QUIZ

1. Transmission lines, properly matched in impedance to the source and load, will have no signal loss.

   a. (   ) true
   b. (   ) false

2. The impedance of a TV transmission line depends on several factors. Which one of the following is not one of those factors?

   a. (   ) diameter
   b. (   ) distance between conductors
   c. (   ) dielectric or insulator material
   d. (   ) length of wire

3. The impedance of a length of transmission line determines the signal loss per 100 feet.

    a. (   ) true
    b. (   ) false

4. A smaller diameter conductor, used in a transmission line, will cause a greater signal loss.

    a. (   ) true
    b. (   ) false

5. Which of the following transmission lines is nearer to 0.4 of an inch in diameter?

    a. (   ) RG-213
    b. (   ) RG-59
    c. (   ) RG-6

Chapter 6

Splitters, Taps, Attenuators, and Traps

In the early years of TV broadcasting, splitters were not a common hardware item for antenna installers. Just about all the installations were composed of a single rooftop antenna and one transmission wire terminating at the TV set. Later some people got a second TV and had another antenna installed to feed it. Still others installed a second set and simply spliced another length of twin-lead onto the original downlead, or attached a second flat-lead to the antenna terminals of the first set, to supply the second set.

SPLICING THE TWIN-LEAD

Splicing transmission line is not a good idea. The line acts as a load to the antenna and an input circuit for the TV set. The impedance between one side of the wire and the other is 300 ohms. Paralleling another piece of twin-lead is like connecting two resistors in parallel. Two equal resistances connected in parallel halves the original resistance, or in this case, the impedance. The wrong impedance causes wasted signal power and reflected energy. Sometimes the mismatch in impedance is not noticeable and the splice works satisfactorily. On occasion equally-spaced, diminishing-intensity ghosts, or ringing, are the result. The mismatch might merely cause what appears to be a smeary picture. Obviously, splices of this type are to be avoided. To attempt to straighten out an antenna system that is performing poorly, one of the first orders of business is to eliminate mistakes like this.

As more homes started getting multiple TV sets, manufacturers began making splitters. Two-way splitters and four-way splitters are the most common. Inside the two-

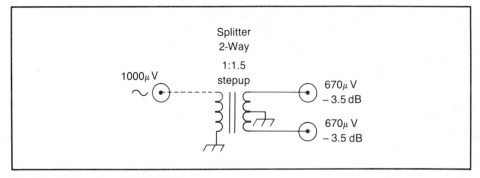

Fig. 6-1. Schematic of 2-way splitter.

way splitter is a small transformer, usually able to pass 54 to 900 megahertz, that divides the signal into two separate output. Figure 6-1 is the schematic for a 2-way splitter.

The two-way splitter is the basis for all other types of splitters. In other words, all splitters are made up of combinations of 2-way splitters. A 4-way is one 2-way, feeding 2 more 2-ways. An 8-way splitter is a 2-way, feeding two 2-ways, that are feeding four 2-ways. If you understand a 2-way splitter, you can easily figure out 3-, 4-, 8-, and 16-way splitters. Because splitters are used in just about every signal distribution system, it is important to know everything about them. Figure 6-2 is a schematic of a 4-way splitter.

You know now that any splitter is simply a combination of 2-way splitter elements. What else is there to know about these small hardware items? There are several other important things to know. First, you need to know what loss there is through a splitter. Why? Because that signal loss can reduce a perfectly good TV picture to a marginal one. You need to know just what to expect when installing a splitter.

A good example is a personal experience I had. In a deep fringe location (60 miles from the prime broadcast stations), at a favorable elevation, the customer had been using a 2-bay, conical antenna mounted on the gable end of the house. From this antenna, the flat-lead dropped 25 feet directly to a window and right inside to the TV (30 feet of wire total).

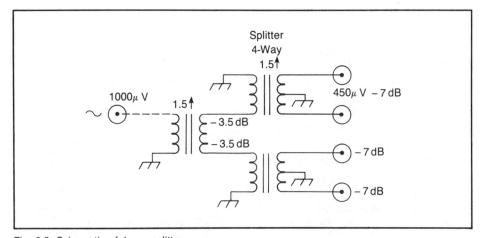

Fig. 6-2. Schematic of 4-way splitter.

34

The customer had adequate vhf reception on the four available vhf stations. The rf signal-strength meter measured from −10 dB to 0 dB. Obviously a 17 dB preamplifier would have made an improvement, and a rotor would have added the ability to pick up from 3 to 7 additional channels. The customer was convinced that her pictures were fine and that she didn't need anything else. Just put up a new antenna with uhf elements and run the flat lead, also run another flat lead to a bedroom set at the opposite end of the house 65 or 70 feet away!

Another possibility was to mount the new antenna on the chimney at the far end of the house. This would have solved the long-wire run problem to the bedroom set, but created one back at the living room set. The reason for considering this option was to mount the antenna in a higher, neater, and more secure location.

Splitting the signal would have introduced a 3.5 dB loss to both TV sets. The loss is even greater than 3.5 dB at uhf frequencies. Without question uhf channels should have been added to the system; however, just this 3.5 dB splitter loss would have caused the signal to be −13.5 dB to −3.5 dB. What had been a marginally good picture (just a slightly busy background on two of the channels and a class A quality picture on the other 3 vhf channels) now would be slightly worse and not satisfactory. By adding 2 dB more wire loss to the run for the second set, the resultant signal was not going to be satisfactory on any channel.

To have proceeded with such a job, attempting to extend and split a system that was already borderline so far as signal is concerned, would have resulted in unsatisfactory picture quality on both sets and a guaranteed irate customer. If the suggestions for an amplifier, higher-gain U/V antenna and a rotor were not heeded, I was fighting a no-win battle. It is better to turn down the job if you are not able to sell what is required. Not knowing line and splitter losses can prevent you from understanding a situation and predicting results.

A value of 3.5 dBmV is the nominal loss figure attributed to splitters, whether they be for twin-lead, or coax. You will find that some splitters are marked as having 4 dB of loss. Since this loss varies with frequency, just like wire losses, you normally consider the "worst" case when figuring a system, knowing that the lower channels and shorter runs will be better.

SPLITTER LOSS

You might ask the question: "Why is the loss 3.5 dB?" Shouldn't it be 6 dB, or ½ of the signal? It would seem that a 2-way splitter would leave only ½ of the original input signal at either output. After all, if you lose only 3.5 dB, or about 34 percent, doesn't this mean that the combined voltages out of both outputs add up to a greater amount than the input?

Because the splitter is concerned only with signal *voltage* levels (as opposed to current or power) there is a step-up transformer action that actually does produce more total voltage in the two separate output ports than was available at the input. This is important and is a definite *plus* in system operation.

The term *loss* is sometimes thought of as a measure of the splitter's quality by novices. Perhaps some other term would better describe the splitter separation of the input signal. It is sometimes called the *isolation* loss. Either way, it is a term that indicates a loss. Actually the splitter does have a loss of power, but since we aren't dealing in power, only signal voltages, the splitter actually appears, and does have voltage *gain*. To one single output port of a 2-way splitter, the voltage level will be reduced 3.5 dB from the input signal level. This is the reason the splitter is said to have a loss. You can figure 3.5 dB loss for any 2-way splitter.

After you have used splitters for a short time, you will notice there are 3-way splitters also. Why? Because three TV sets became common in many households, there was a demand for 3-way splitters. You may ask "If the splitters are merely combinations of 2-way splitters, isn't a 3-way really a 4-way with one outlet internally terminated?" That's nearly right. Actually, the 3-way is one 2-way with one of the ports used as one of the 3-way ports (−3.5 dB loss). The other 2-way port goes into second 2-way. Both of these output ports are used as the 2nd and 3rd ports, at 7 dB loss. By making 3-ways like this, rather than using a 4-way, one output is at a higher level and could conceivably be used for the longest run, or perhaps for the most-used TV set, if the system was operating marginally on some stations.

Recapping

A brief review of the important points regarding splitters:

- Splitters are always combinations of 2-way splitters.
- Splitters have a nominal 3.5 dB loss per split output port.
- 4-way splitters have a 7 dB loss per outlet port.
- The splitter loss is sometimes called isolation loss.

DC PASSIVE SPLITTERS

One last parameter of a splitter is whether it is dc passive or not. Some splitters will not pass dc or low frequency ac to either output port. Some are made where both outputs, or all outputs, are dc passive. Some are made where all but one are not passive and only one is passive.

Sometimes it is convenient to split the amplified signal coming down from the pre-amplifier, before it enters the power supply section. For instance, the power supply section might best be located in the family room, 50 to 100 feet from the antenna while a second outlet is needed only a few feet from the antenna in the bedroom. Rather than running a separate wire all the way back up to the bedroom, split the download at the bedroom entry. The splitter should be passive on the side going to the family room so that the AC power, supplied by a small 14 Vac transformer in the power supply, can reach the amplifier through the downlead.

If you should be in a position where you have nothing but passive both-way splitters, you can still use them, but insert a dc block in the leads not powering the amplifier. Otherwise, the power supply (14 Vac) can be shorted out by the bandsplitter attached to any other TV set. These bandsplitters usually are short circuits to 14 volts ac, or dc. If the power supply gets hot and burns out, or if the reception degenerates, it is wise to check for this possibility.

Splitters Can't Be Used Everywhere

Now that you know about splitters you might feel that you can hook up any number of TV sets merely by dividing the available signal properly. The problem is that if you use a 4-way splitter, you have reduced the original signal by a little more than half. That isn't bad, but suppose you have 8 sets? (a college dorm, an apartment building, classrooms). If you split again you have reduced the signal, reduced by 7 dB already, by another 3.5 dB. Now you have lost 10.5 dB, plus other losses. Other losses are wire losses of 2 to 3 dB per 100 feet for vhf channels and 6 to 9 dB for uhf! If you started out with a +5 dB signal and lost 12 to 16 dB, you can find that you now have a class B or worse picture. Adding a stronger front-end amplifier or booster might help the end-of-the-line sets, but it could produce too large a signal near the front end. Maybe there is a better way than simply using splitters everywhere.

TAPS

Taps are similar to splitters. (See Fig. 6-3.) Notice that the input is tied directly to the feedthrough output. The output is merely the trunk line, or a continuation of the input. The tap is also an output. The tap-off is connected to the TV set you are feeding. Obviously, if the trunk line input is connected to the trunk line output, there should be very little loss with respect to the trunk line signal. In Fig. 6-3 we are using a −17 dB tap. This means that whatever signal voltage is fed through from input to output, the tap to the TV set will be lower by 17 dB. Only about 13 percent of the original input signal can get through the 1 nanofarad capacitor and the 470 ohm resistor to supply the set. The beauty of using a tap is that if you started out with a +20 dB signal, you could supply quite a few taps with a fine signal before your trunk line lost an appreciable portion of its original voltage level. (+20 dB −17 dB = +3 dB in this case, for the sets connected to the first tap, and only slightly less at the second and third taps.)

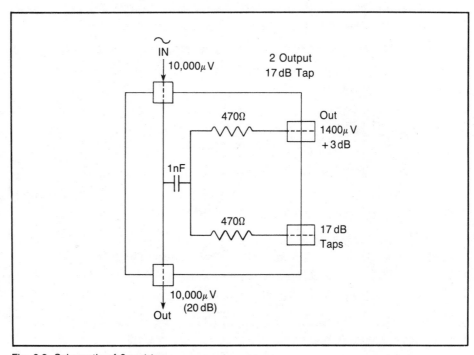

Fig. 6-3. Schematic of 2-port tap.

Insertion Loss

Because each fitting, or "F" connector, has a slight loss, and because the tap feed-through has some sharp bends in the wire, each tap has a little loss from trunk line input to trunk line output. This varies, but a −17 dB tap will have about 0.5 dB or loss from input to output. (Remember the tap-off ports still have −17 dB loss.) If you use 6 taps, the initial +20 dB signal will have been reduced by 3 dB at the 6th tap, or effectively at the 17 dB tap-offs of the 6th tap, 0 dB is still sufficient for a class A picture.

Using the term *insertion loss* is very confusing. To me, that should mean the loss required to insert the signal, from the trunk line to the TV set being supplied by the tap. However, the feedthrough loss will be called insertion loss and is about ½ dB for a 17 dB tap. You do need to account for the insertion loss in any sizable MATV system. If you will remember the loss is 0.5 dB for a 17 dB tap, it is slightly less for a 24 dB tap and a little more for a 12 dB tap, you will have a "handle" on how taps work.

Why Different Tap-Off Losses?

Figure 6-4 shows a variable tap. Again, the input is a direct wire or PC trace to the trunk line output. Instead of a 1000 picofarad capacitor to the isolation resistor (470 ohms in Fig. 6-3), we use a rheostat, a 500 picofarad capacitor in series with a fixed 180 ohm resistor to feed the primary of a small transformer, and another 500 pF capacitor to the tap-off connector. This circuit allows us to vary the loss from −10 to −25 dB. By using variable taps, as you lose signal due to insertion and wire losses, you can reduce the tap loss, increasing the voltage tapped from the trunk line, producing a proper level to the TV set.

Since variable taps are more expensive than fixed, the fixed variety are more common. Since they are not adjustable, it might seem that as you progressed farther and farther

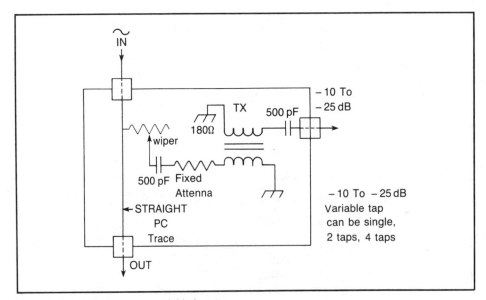

Fig. 6-4. Schematic of 1-port variable-loss tap.

from the original antenna signal, the TV sets would receive less and less signal. What happens is, you start out with perhaps +35 dB of signal, split the signal into two legs, having only 31.5 dB. You use a −24 dB tap on the next few drops down the trunk line. As the insertion losses increase, you drop down to a −17 dB tap, then to −12 dB taps, and if needed near the end to a −6 dB tap. All the sets then receive signals of like amplitude.

The loss between trunk line and TV outlet(s) is called the isolation loss. I prefer to call it the tap-off loss.

Notice that taps are completely isolated from the trunk line, unlike some splitters. This can be helpful should a set owner short out a port, there will be minimal harmful effects on the trunk line. Any unwanted signals fed back into the tap from the TV set outlet, from a computer, a calculator, or an FM radio, will also have little effect on the trunk.

Terminators

Taps are made commonly with 1, 2, 4, or 8 tap-off ports; 2 or 4 are the most common. Any unused port should be terminated with a 72 ohm resistor.

If you now feel you know about splitters, terminators, and taps, you might feel you can handle *any* signal distribution system. You certainly have the major "characteristics" nailed down. But what about attenuators?

ATTENUATORS

Attenuators are merely resistors. They cut down the signal in cases where it is so high that the TV set would overload. In the earlier days of TV, the input signal to the set's antenna terminals had to be close to 1000 microvolts, or automatic gain-control circuit could not keep the locking pulses at the proper level to control vertical and horizontal synchronization. Too large a signal would cause the picture to start rolling or flipping sideways, and the video would "wash out." Modern TV sets have better stability, and can handle signals of 30,000 microvolts (30 dB) and more. Generally, inputs above 30 dB on small MATV systems start introducing some strange problems. It is desirable to keep levels to the TV set at −5 dB to +20 dB.

Different conditions cause different signal problems. Too large a trunk line signal can cause a channel adjacent to the one you want to watch to be amplified to the point where it overlaps and interferes. FM radio stations and adjacent channel TV sound can be amplified too much, causing herringbone patterns all over the screen. In MATV systems, attempting to put different channel signals, with widely varying levels, into a trunk line amplifier will cause the higher level signal to "swamp out" adjacent channels that need the additional amplification. The preferred method is to attempt to equalize the signals coming from the antenna(s) prior to amplifying them, and prior to feeding the trunk line to the tap-offs. One way to equalize the signals is to use attenuators to lower the high-level channels. Obviously, to lower a channel, selectively, you would probably need to use separate antennas for certain channels; if an all-channel antenna is used, you would use a band-separator, or channel separator first.

Variable Attenuators

A most useful piece of test equipment is a 0 to 18 dB attenuator. Carrying one in your tool caddy is wise. Where co-channel, or adjacent channel interference causes a poor quality video, or the TV set is sensitive to any signal variations, placing a variable attenuator in the line to the set can help to solve the problem. With satellite TVRO systems, it is common practice to reduce the receiver *if* level, by inserting a 12 dB attenuator in the *if* line. The attenuators fit easily in the RG-59 and RG-6 transmission lines. The first time you encounter a TVRO signal that is too strong, you will find it looks very much like a weak signal you might not even try an attenuator.

TRAPS

To solve problems with unwanted signal levels or unwanted signals instead of using an attenuator, you could find a trap is a better answer. FM traps are especially useful and quite common. They are merely notch filters that suppress the FM band (88-108 MHz). The TV set has such a trap; it also has traps in the *if* strip for rejecting frequencies other than the 6 MHz *if* passband. If your set is located approximately one block from an FM transmitter (or an offending TV station), and you are attempting to receive a TV station 60 miles away, you can have a problem with interference. The offending signal can be so strong it can still sneak through the traps in the set and cause herringbone patterns. Putting additional FM traps ahead of the set can reduce the effects, while leaving the desired signal alone. There are other traps available to eliminate strong out-of-band transmissions, such as CB radio. Some line amplifiers offer attenuation of any of the 3 TV bands, plus additional FM filtering.

In the early days of TV, before specialized traps were made, technicians would run a piece of twin-lead parallel to the antenna terminals. This short piece of twin-lead would be cut to a particular length, with the outer ends shorted, this stub would act as a resonant cavity, or trap, reducing the signals that it was cut for. A modification on this was a similar lead, with a small trimmer capacitor on the end, rather than a short. The trimmer could be adjusted to maximize the trapping effect.

A final method of eliminating undesirable signals is to reverse the phase of the undesirable signal by passing it through an amplifier stage, then reinsert it into the signal line, thus canceling itself. This is also used in TVRO, where more common methods are unsuccessful.

QUIZ

1. If a splitter divides the original signal, supplying each of two output ports with one-half of the input voltage level, each port will have a 6 dB loss.

   a. ( ) true
   b. ( ) false

2. Splitters are typically listed as having a nominal 3.5 dB loss.

   a. ( ) true
   b. ( ) false

3. If a splitter has an input of 1000 microvolts, each output port should have _____ microvolts of output.

   a. ( ) 1000
   b. ( ) 500
   c. ( ) 250
   d. ( ) 670

4. Since the input of a tap is connected directly to the feedthrough output, there is not feedthrough or insertion loss in a tap.

   a. ( ) true
   b. ( ) false

5. An attenuator is a resistive pad that reduces the signal level by a set dB amount.

   a. ( ) true
   b. ( ) false

6. A TV set can be used to terminate a splitter or tap output port (or feedthrough output port). If a TV set is not used, a/an _____ should be used.

   a. ( ) attenuator
   b. ( ) stub
   c. ( ) isolator
   d. ( ) terminator

7. Herringbone interference patterns on TV screens are usually caused by:

   a. ( ) CB radio transmissions
   b. ( ) FM radio or TV transmissions
   c. ( ) standing waves
   d. ( ) wrong value splitters

8. Splitters and taps would have less losses if they were made of higher quality components.

   a. ( ) true
   b. ( ) false

9. DC blocks are used in coax transmission lines to:

   a. ( ) pass direct current while filtering all AC signals
   b. ( ) pass AC power supply voltage, but prevent any DC from reaching the preamplifier
   c. ( ) prevent AC voltage from reaching the preamplifier
   d. ( ) prevent AC power supply voltage from being shorted by a balun or bandsplitter.

10. Mast-mounted preamplifiers are supplied 14 volts AC through the same coax or twin-lead transmission line that delivers the TV signal from the preamp to the preamp power supply, and thence to the TV set or distribution system.

    a. ( ) true
    b. ( ) false

Chapter 7

Preamplifiers,
Boosters, Line Amplifiers

The introduction of low-cost, transistorized, mast-mounted, rf signal amplifiers was a major improvement in the TV reception business, prior to that vacuum tube technology was used. Some felt the increased gain produced by the tubes was just about offset by the noise generated by the tube circuitry. Tubes also had a shorter life than transistors. The tube-type amplifiers couldn't be mounted at the antenna itself. Mast mounting of the transistor preamplifiers meant the shortest amount of lead wire could be used, providing the maximum signal from the antenna elements. There was a reduction in the signal-to-noise ratio on the downlead from the amplifier to its power supply.

PREAMPLIFIERS

Figure 7-1 shows the block diagram of a M/A-COM LA-10 preamplifier. Note that this device has a separate power supply section, in addition to the actual amplifier circuitry.

The power supply section performs two functions: First, it contains a 110 Vac to 14 Vac stepdown power transformer. Second, it isolates the signal output from the 14 Vac line voltage. The power-supply transformer secondary sends 14 Vac up to the mast-mounted amplifier. The PC board-mounted power-supply circuit amplifier rectifies and filters the 14 Vac and uses the resulting dc voltage to power the transistors that do the amplification. On the same downlead that transfers the 14 volts up to the preamp section, the amplified signal is impressed and brought down to the power supply section. Nothing happens to the signal in the power supply, except that the signal is passed through a small capacitor to the output connected to a TV, or MATV head-end. The capacitor is too low in value to pass the 60 cycle line frequency, but easily passes the rf signals.

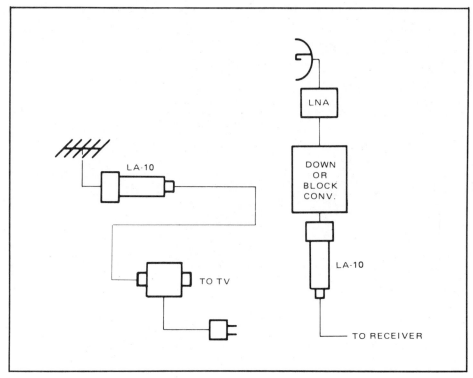

Fig. 7-1. Drawing of M/A-COM LA-10 in-line preamplifier.

It is not necessary to connect the output port of the power supply to the MATV system or TV set, although it is the usual procedure. Sometimes it is more practical to connect a TV somewhere between the mast-mounted amplifier and the power supply. Since the same signals are present on the downlead, as at the output of the power supply section. Figure 7-2 is a typical preamplifier circuit schematic.

PICKING OFF SIGNAL FROM THE DOWNLEAD

In order to take signal from the downlead, a splitter must be used. How else could you power the amplifier section, plus a TV set along the way? If you use a non-passive splitter, you will be blocking the 14 Vac power to the amplifier, that won't work. If you use a splitter that passes dc and low frequency ac on both sides, it will work. It will work only until you hook the intended signal lead to a bandsplitter, balun, terminator, or other device that uses a very low impedance transformer to couple the signal. U/V bandsplitters are the worst culprits because they act as a dead short to the 14 Vac with just the splitter coils isolating the signal from the 14 volts. Such short circuits will burn out power-supply transformers and make systems inoperative. Novice reception technicians often hook up systems with this fault and assume that the poor quality pictures are due to the poor location. In effect the amplifier is not working at all, and only a small portion of the signal is feeding through the amplifier circuitry. Usually, the uhf channels will feed through best, leading to the conclusion that the amplifier is in fact, working.

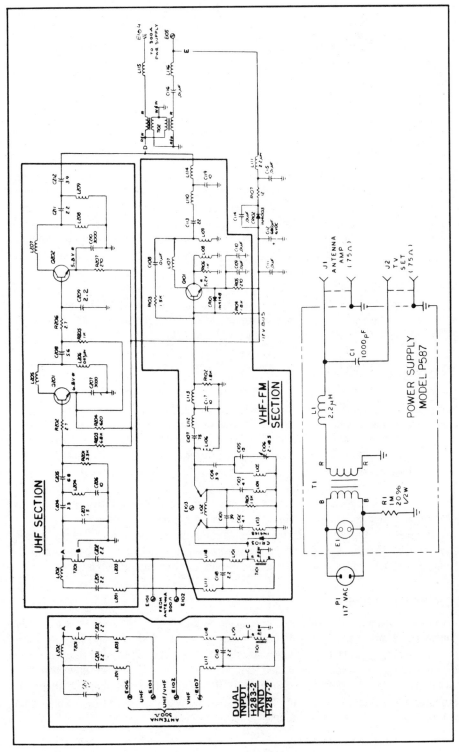

Fig. 7-2. Schematic of G.I. preamplifier.

Fig. 7-3. Drawing of dc block.

DC BLOCK

The solution to the passive splitter problem is to install a dc block (See Fig. 7-3) in the signal output leg of the splitter. This way any device can be hooked to the lead, and no effect will be seen on the 14 volt power. A better solution is to carry splitters that are dc passive on only one port. This is the port used to connect the power supply section. One can also use bandsplitters that are not dc passive, or hook the coax lead directly to a TV set that is not passive. This will be a temporary solution, because eventually someone will hook up a dc-shorted bandsplitter and short the line.

The amplifier section circuitry is composed of one or more transistors. The amplifier's job is to increase the signals. The most common preamplifier gain is 17 dB. This means the original signal from the antenna elements will be increased to about 8 times its input size. A −30 dB signal will be increased to −13 dB by a 17 dB amplifier. Most U/V amplifiers provide 2 or 3 dB less output for uhf.

U/V PREAMPLIFIERS

Since uhf stations are ordinarily more difficult to receive given the same distances and conditions, most preamplifier manufacturers make a model with separate uhf and vhf twin-lead antenna inputs. This allows the use of a separate higher-gain uhf antenna. It allows for a higher gain vhf antenna too, if that is the needed solution. With separate inputs, the uhf and vhf antennas can be aimed in different directions, if necessary. These U/V separate-input preamplifiers come optionally with uhf gain 6 to 8 dB higher than vhf. At 50 miles from the broadcast station, it is typical the uhf signals are down 10 to 15 dB from vhf. With 6 dB more of uhf amplifier gain, and up to a 6 dB improvement in antenna selection for uhf, the signals can be made similar in strength.

While the preamp schematic might at first seem complicated, it is actually a straightforward wide-band amplifier, usually containing a switchable FM trap. Some offer tunable FM traps to further aid the technician in overcoming interference.

BOOSTER AMPLIFIERS

Booster amplifiers are similar to preamplifiers, and can be obtained with a variety of gain specifications. Technicians can use boosters in tandem with preamps for conditions where the preamp is not quite sufficient to produce a snow-free picture, or, where several new room outlets are going to be added. Suppose you intend to replace one drop with a 4-way splitter leading to other rooms. A 10 dB booster amp will make up the 7 dB splitter loss, plus 3 more dB for line losses. This can provide each of the four outlets with the same signal the original single set was receiving.

LINE AMPLIFIERS

Line amplifiers are used in larger systems, they are usually higher in price and higher in gain than most homeowners care to have, but some are occasionally found in homes. Most often you will encounter them in apartments, mobile home parks, schools, and so forth.

Line amplifiers typically have gains of 35 dB to 64 dB.

Line amplifiers often have separate band-level adjustments as well as FM traps. In using any signal amplifier care must be taken to match the amplifier to the job. The input circuit will not accept all signal level from −40 dB to +40 dB. A preamp or booster can be swamped or overloaded by a 20 dB signal (0.01 volts), while line amplifiers will accept up to 35 dB (0.056 volts ac).

The use of amplifiers is somewhat of a science in itself. The MATV technician should experiment with all types to learn the capabilities and proper applications for each type.

SMATV AMPS

Cable and large MATV systems need to use many available channel positions. Individual channel separation is another facet of the installation. After separating a channel signal, it can be reduced or amplified, so that it and all other channels to be used, are of equal size. Other processing might also be done, such as trapping out an overly strong adjacent channel or FM radio station signal. Once the signal is processed, it and all other signals are combined into the trunk line(s).

The SMATV system might utilize satellite signals from several TVRO stations, off-air uhf and vhf channels, and locally generated video, perhaps from a VCR or surveillance camera.

There are several reasons for converting some channels to a different frequency or channel position. For instance, if channel 6 is extremely strong in your area, the signal would be picked up on the unshielded input wiring on the TV set(s) in some cases. This would cause a forward ghost in the picture. Converting channel 6 to an unused uhf location (or an unused vhf slot) would eliminate the problem.

In some cases the wiring used in an MATV system is old and only usable on vhf due to the high loss it has at ultra high frequencies. If there are open vhf station slots, the uhf stations can be converted down to the vhf locations, solving the problem. Other problems can be solved by converting and separating bands, or individual channels. There are several manufacturers supplying head-end hardware to solve practically any problem. Blonder Tongue, Channel Master, Winegard, M/A-COM, and General Instruments are old reliable companies in this business. They supply catalogs that contain specifications on each product. Reception technicians can request catalogs from their electronic parts distributors, or direct from the manufacturer. Each contains information about MATV and antenna equipment that is educational and invaluable to the SAM technician.

Their addresses are:

Blonder Tongue
1 Jake Brown Rd.
Old Bridge, NJ 08857

M/A-COM Industrial
8240 Haskell Ave.
Van Nuys, CA 91406

Channel Master
PO Bx 1416
Smithfield, NC 27577

General Instruments
RF Systems Div.
1 Taco St
Sherburne, NY 13460

Winegard Co
3000 Kirkwood St
Burlington, IA 52601

QUIZ

1. Mast-mounted preamplifiers normally are powered by:

    a. ( ) solar cells
    b. ( ) rectification of the rf signal energy
    c. ( ) batteries
    d. ( ) 14 volts ac from the power supply section

2. To use a separate U/V input mast-mounted preamplifier with a broadband single output U/V antenna . . .

    a. ( ) connect the antenna lead wire to the vhf input
    b. ( ) connect the antenna lead wire to the uhf input
    c. ( ) uhf or vhf can be utilized, but not both
    d. ( ) a band separator must be used, with separate wire to each preamp input

3. The optimum signal level for a TV set will be:

    a. ( ) −20 dB
    b. ( ) +20 dB
    c. ( ) −40 dB
    d. ( ) +40 dB

4. Which of the following might have a 60 dB output level?

    a. ( ) preamplifier
    b. ( ) booster amplifier
    c. ( ) line amplifier

5. The output of a preamplifier should never be taken from the line between the power supply and the mast-mounted amplifier section.

   a. ( ) true
   b. ( ) false

6. What type of interference might be caused by off-air TV channels 2 and 4, plus a satellite dish operating on channel 3?

   a. ( ) co-channel interference
   b. ( ) adjacent channel interference
   c. ( ) ghosts
   d. ( ) ringing caused by standing waves

7. One solution to the problem in question 6 could be to:

   a. ( ) convert channel 4 to an unused uhf channel position
   b. ( ) convert channel 2 to an unused uhf channel position
   c. ( ) convert channel 3 to an unused uhf channel position

8. The off-air channels to a home perform as follows at the output of the preamplifier: channel 4 = +5 dB; channel 10 = 0 dB; channel 13 = −8 dB; channel 30 = 0 dB. The most economical improvement would be to . . .

   a. ( ) add a 35 dB line amplifier
   b. ( ) add a 10 dB booster amplifier
   c. ( ) increase the transmission line wire size
   d. ( ) add a converter to place channel 13 on an unused uhf channel

9. The off-air channels to a home perform as follows: channel 2 = 35 dB; channel 8 = −10 dB; channel 30 = +4 dB. There is a need to split the signal 8 ways, what is the most economical method?

   a. ( ) use individual channel separators, then individual channel amplifiers to boost channel 8 and channel 30
   b. ( ) use a 3-band line amplifier, adding more power to channel 8 and channel 30. Reduce channel 2 output. Connect output of amp to two 4-way 6 dB tap-offs
   c. ( ) separate the uhf channel from the vhf with a U/V separator/joiner. Separate channel 2 from channel 8 with a vhf Hi/Lo separator/joiner. Attenuate the Lo vhf (channel 2) with an in-line 20 dB pad. Combine channel 2 and 8 into a Hi/Lo joiner/separator. Combine the uhf and vhf channels through a U/V joiner/separator. Put entire U/V output into a 24 dB booster amplifier and then into an 8-way splitter.

Chapter 8

Small MATV Systems

In the 1950s one TV was considered a luxury. People now realize that TV is not just an entertainment toy; it is a communications device. They rely on TV for weather alerts, news, major sports events, and educational shows. It makes sense to add TVs to other areas of the house, such as the bedroom, the kitchen, the den, or the workshop. Rabbit-ear antennas and improper twin-lead splices have worked for some, in hooking up the second and third sets. Others have bought booster amplifiers to provide the larger signal needed when the antenna signals are divided up among several TVs. Large houses often require long cable runs. More and more builders are including pre-wired antenna systems in the construction plans. Prewiring places the twin-lead or coax in the walls with neat outlet plates for attaching a TV in any room desired.

In the early years of TV, prewiring was often done by electricians who had little or no antenna background. Some of the reasons for poor results are:

- No signal amplifier
- vhf-only components
- 17 dB tap-offs used everywhere
- Excess twin-lead and coax
- Twin-lead laid over metal ducts, along masonry walls, or stapled in place with wire staples.
- Poor quality twin-lead and coax
- Non-terminated outlets

These are a lot of problems for a simple home TV antenna system. Most home MATV systems have one or more of the above problems. Some have all of them! Some have additional problems with the antenna and rotor, which may not have been considered in the original house pre-wiring plans. Because homeowners had no way to tell whether something was not right with the system, most simply put up with what they got. They might blame the poor quality reception on location. One or two local stations would have good reception, even on the poor system, so those were the only ones watched. Others were concluded to be too far away or in the wrong direction.

The desire to receive more channels and the exposure to video discs, VCRs, cable, and satellite signals has put pressure on people to improve their reception. If the pictures from these new sources can be perfect, why can't the local channels?

Millions of inadequate single-family-dwelling MATV systems need repair, replacement, or improvement. The public really needs to be educated to understand the reception standards for their areas. Since cable companies and the promoters of other forms of video media have built an industry enhanced by the existence of poor TV reception, we can't really expect them to make any suggestions for off-air TV reception improvement. They love it the way it is. The TV networks and local stations seem unconcerned, or perhaps they feel promoting better reception to their remaining viewers is too large a task.

THE NEED FOR SAM TECHNICIANS

Those who are in the reception business: TV, antenna, and satellite dealers, and technicians are left to sell the public on improved antenna and MATV systems. The public is willing. The programming is available, and the hardware is made by several manufacturers to do the job right. All that is needed is more technicians willing to troubleshoot bad systems, and to correct them.

Starting from a basic antenna hookup, we will show several examples of systems that need some improvement. We will list one or more possible solutions to the problem and the results expected:

Case #1

Description:
Antenna: Conical
Wiring: Twin-lead, 50 feet
Distance from broadcast station: 20 miles
Available channels: 4 vhf, 3 uhf
Distribution: 1 TV in living room
Signal levels: channel 4 = -5 dB
channel 6 = $+10$ dB
channel 8 = $+5$ dB
channel 13 = 0 dB

Improvement Suggestion:
Replace conical antenna with U/V short-range antenna. Replace lead wire with coaxial cable. Install U/V bandsplitter on the TV set. (See Fig. 8-1.)

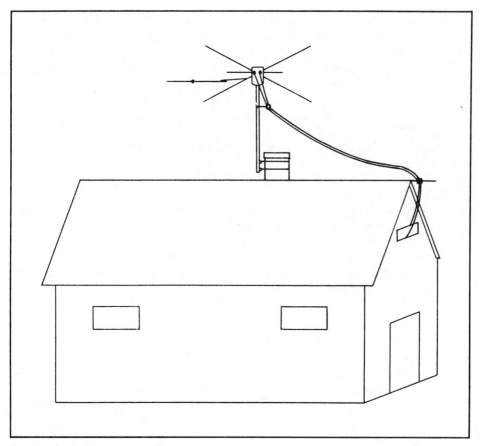

Fig. 8-1. Conical antenna on a house.

Case #2

Description:
Antenna: U/V medium range (6 to 10 dB gain)
Wire: Twin-lead 50 feet
Distance from broadcast station: 20 miles
Available channels: 4, 6, 8, 13, and 3 uhf from the East.
Levels: channel 4 = −10 dB
channel 6 = +10 dB
channel 8 = +10 dB
channel 13 = +5 dB
20, 40, 59 = −10 dB
Distribution: One TV in living room

Improvement Suggestion:
    Replace twin-lead with coax. Install a 12 to 17 dB booster amplifier. (See Fig. 8-2.)

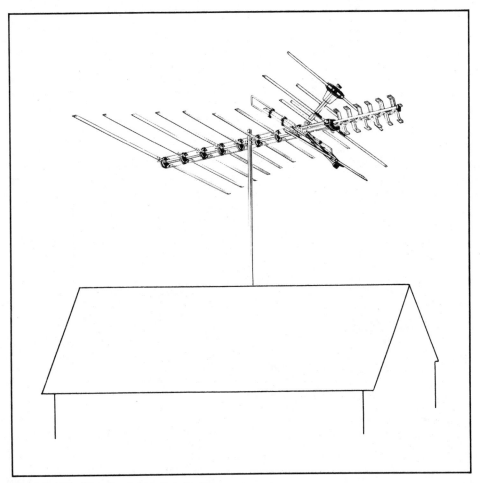

Fig. 8-2. U/V all-band antenna on a house.

Case #3

Description:
Antenna: Conical
Wire: Twin-lead
Distance from broadcast station: 20 miles
Available channels: 4, 6, 8, 13, 20, 40, 59
Reception: Ghosts on all channels. Signal strength is +5 dB or better on all channels. Ghosts caused by nearby power lines and towers.

Improvement Suggestions:
Install a rotor with a new antenna that has a very narrow polar pattern, and a high front-to-back ratio. (See Fig. 8-3.)

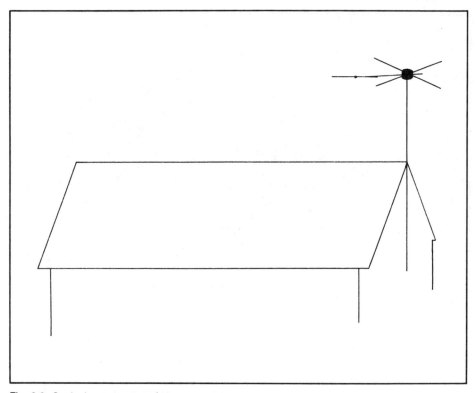

Fig. 8-3. Conical antenna on a house.

Case #4

Description:
Antenna: Medium range U/V
Wire: Twin-lead to a 4-way splitter in the attic
Distance to broadcast station: 40 miles
Available channels: 4 vhf, 3 uhf
Reception: channel 6 = −2 dB, others worse
uhf = −20 dB

Improvement Suggestion:

Install a 17 dB or larger preamplifier. Install coax to power supply plugged into an attic outlet, then run the output, through a balun to the existing 4-way splitter. (See Fig. 8-4.)

Results should be: channel 6 = +14 dB
Other vhfs to the + dB range
uhfs to −5 dB or better. (With a 30 dB preamplifier)
The uhf would be +10 dB or better.

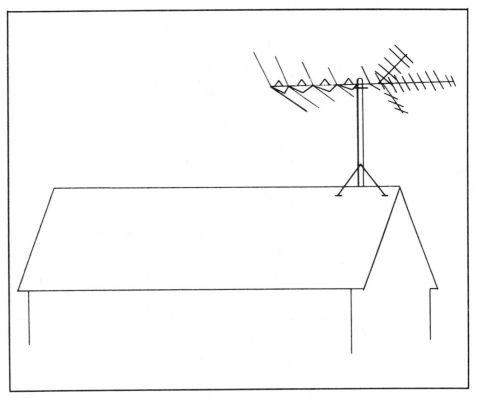

Fig. 8-4. U/V all band antenna on a house.

Case #5

Description:

Same as case #4 except coax had been used instead of twin-lead. Three outlets are all that is ever needed, but there is a 4-way splitter in the attic. A long run to the family room has left uhf at −26 dB and vhf channels in the negative range. The household is primarily interested in uhf channels.

Improvement Suggestion:

Add higher gain uhf antenna section to present U/V antenna. Use a 17 dB for vhf and a 24 dB uhf preamplifier. Replace 4-way splitter with a 3-way, using the 3.5 dB output port for the long family room run. Results should change to:

Channel 6 = +18 dB

Uhf channels will be 3.5 dB better in the family room, due to splitter gain, 6 dB better due to new higher antenna gain, and 7 dB better due to separate input (24 dB) amplifier.

Vhf channels should be at +6 dB or better. (See Fig. 8-5.)

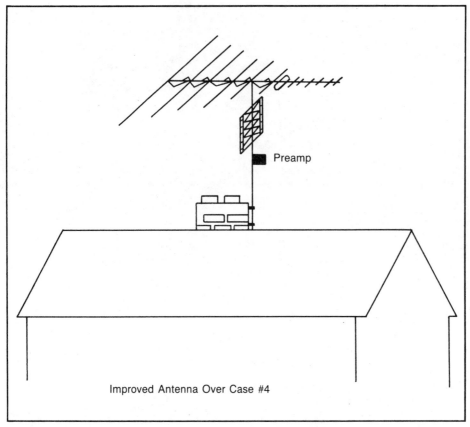

Improved Antenna Over Case #4

Fig. 8-5. U/V and uhf 4-bay with preamplifier mounted on a chimney.

Case #6

Description:
Antenna: double-bay conical
Wire: twin-lead, 300-ohm
Distance from broadcast station: 20 miles
Channels available: 4, 6, 8, 13, 20, 40, 59
Reception: poor, vhf only, signal -6.5 to -10 dB

Improvement Suggestion:

Install separate uhf and vhf antennas, use a 17/24U preamplifier. Use a 2-way passive splitter, and 17, and 12 dB tap-offs. (See Figs. 8-6 to 8-9.)

Lower cost alternative:

Install preamplifier on the existing conical antenna and add a uhf section antenna; use coax cable with 17, and 12 dB tap-offs.

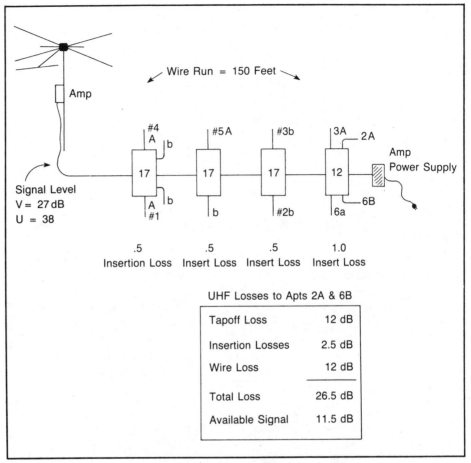

Fig. 8-6. Multiple family dwelling antenna system using conical and preamplifier, also signal level insert is shown.

QUIZ

1. While the Indiana TV-Radio License Examiners Board rules state that an MATV system is any TV antenna system having 5 or more outlets, there is no generally accepted dividing line between an antenna system having multiple outlets, and an MATV system.

   a. ( ) true
   b. ( ) false

2. A TV antenna, connected to a 2-way splitter, then to two TV outlets, would generally not be considered an MATV system.

   a. ( ) true
   b. ( ) false

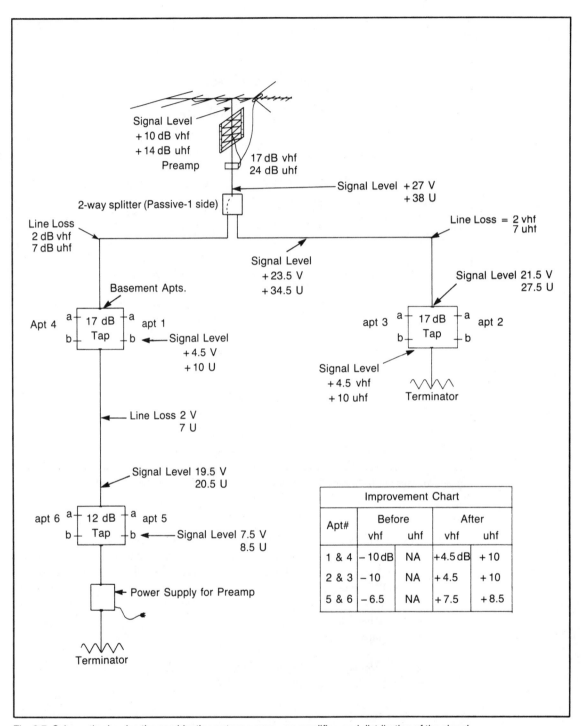

Fig. 8-7. Schematic showing the combination antenna array, preamplifier, and distribution of the signal.

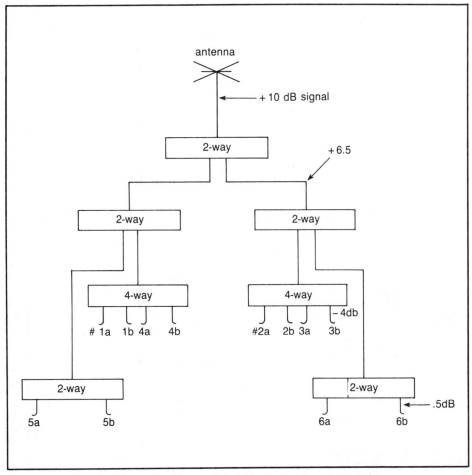

Fig. 8-8. Drawing showing apartments as originally hooked up to splitters only.

3. A TV antenna, using a preamplifier, then divided to serve 7 rooms, would be classified as a small MATV system.

   a. ( ) true
   b. ( ) false

4. If an antenna produces a 10,000 microvolt signal, and the wiring, splitter and tap-off losses to the farthest port are 20 dB, the resultant signal at that port will be approximately:

   a. ( ) 10,000 microvolts
   b. ( ) 10 dB
   c. ( ) −10 dB
   d. ( ) 0 dB

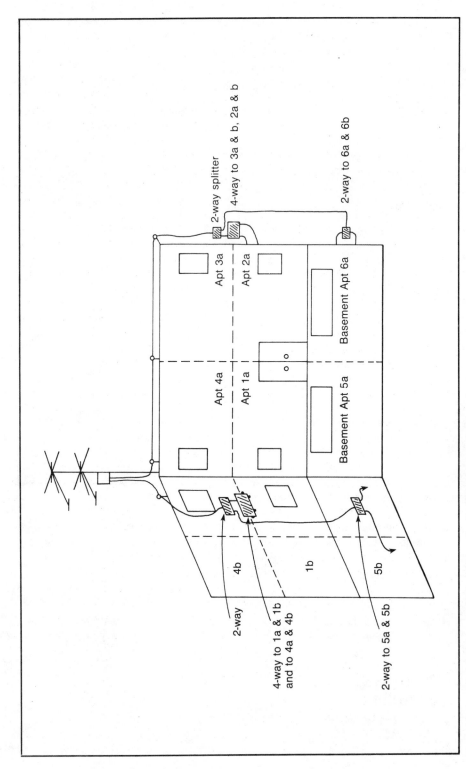

Fig. 8-9. Apartment wiring diagram showing the hookup using splitters only.

2-way splitter

4-way to 3a & b, 2a & b

2-way to 6a & 6b

Apt 3a

Apt 2a

Apt 4a

Apt 1a

Basement Apt 6a

Basement Apt 5a

2-way

4-way to 1a & 1b
and to 4a & 4b

4b

1b

5b

2-way to 5a & 5b

5. Twin-lead distribution system wiring theoretically has less loss per 100 feet than coaxial cable; however, in actual practice it is a poor choice. Some detrimental aspects of twin-lead are:

    a. ( ) splitters are not available
    b. ( ) preamps are not made to accept twin-lead, without adapters.
    c. ( ) antennas are not made to accept 300 ohm lead wire
    d. ( ) direct pickup and increased loss due to moisture

6. An SMATV system is a MATV system that also includes several:

    a. ( ) satellite channels
    b. ( ) splitters
    c. ( ) separate buildings or structures
    d. ( ) subsystems

7. The higher the isolation value of a signal tap-off, the _____ the insertion loss

    a. ( ) higher
    b. ( ) lower

8. Due to taps having high isolation between trunk line and port, 75 ohm resistive terminators are not necessary.

    a. ( ) true
    b. ( ) false

9. Generally speaking, a MATV system supplying 50 dB of signal to the tap-off or splitter ports, is preferred over one supplying only 10 dB.

    a. ( ) true
    b. ( ) false

10. Starting with 50 dB of signal at the MATV head end, and supplying 25 units located in a row and spaced 50 feet apart, it would be best to use: (to each outlet)

    a. ( ) 8-way splitters
    b. ( ) 17 dB taps
    c. ( ) 24 dB taps

Chapter 9

TVRO Principles

The ultimate form of communications is a name given to satellite communications technology. From an idea published by Arthur Clarke in 1945, aided by World War II, subsequent rocket technology, and made practical by government space programs such as President Kennedy's "Landing a man on the moon" program, today the world is rushing to equip itself to utilize satellite communications. Where land-based TV transmission is limited by terrain, TVRO (Television Receive Only) is not. Virtually everyone in the world can, or shortly will be able to receive satellite TV. Figure 9-1 shows the location of major TV satellites in the Clarke Belt.

The home TVRO industry wasn't planned, it just happened. Programmers like HBO and the TV networks found satellite relaying more economical and dependable than land microwave, and telephone lines. RCA launched its own SATCOM satellites, and Western Union put up WESTARs, Hughes and others followed to where now, the problem is that the available "parking space" for more than 20 satellites in the equatorial geosynchronous belt is limited.

Electronics experimenters started building TVRO systems, and simultaneously the programming available grew to dozens of channels. Manufacturers recognized the emerging home market, as well as the market for cable and broadcasting equipment. Names like Chaparral, Drake, Uniden, Birdview, and Janeil became well known within the TVRO industry. Electronics product distributors jumped on the bandwagon and new companies sprang up to supply the growing public desire to bring the satellite programming into the home.

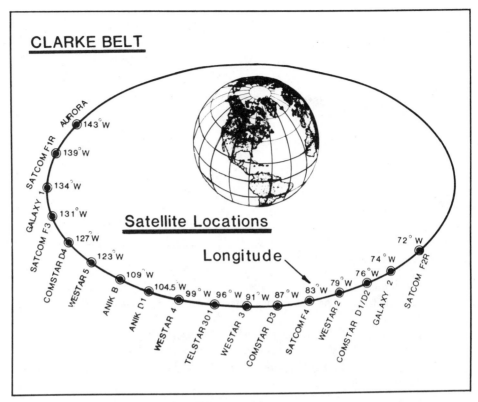

Fig. 9-1. Clarke Belt positioning of satellites.

Consumer electronics technicians and dealers for the most part weren't the pioneers in the dish business. Most TV-audio, and electronic service businesses took a wait-and-see attitude. Only a small percentage ever eventually took the plunge into TVRO sales and service. Dish installers seemed to come from many non-related areas in addition to a sizable number of especially rural, TV-radio service firms. Many satellite retailers are part-timers. Farm equipment, REMCs, garage and swimming pool sales firms, cable companies, service station operators, and others decided to get into the dish business. The satellite system was a big-ticket item and the potential profits were enticing.

Even before the 1986 slowdown, caused by scrambling of the movie channels, few retail dish firms lasted for as long as one year in the business. Not only the opportunists, but some well qualified electronics dealers and technicians found the TVRO business very difficult, and were forced to get out. The reasons for the high mortality rate seemed at least in part to be:

- What appears to be a needed business investment for 1 or 2 dish systems, and a small amount of tools and equipment, (perhaps $5,000 to $10,000) quickly turns into several times that amount.
- While the installation, hookup, and aiming of a system seems simple, it frequently becomes a difficult, time-consuming, and expensive task. Each job introduces the installer to new, and often near-insurmountable problems.

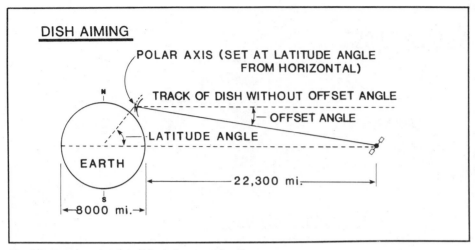

Fig. 9-2. Drawing showing dish aiming considerations.

- Distributor and manufacturer mistakes are practically always paid for by the dealer. Equipment failure, unclear instructions for assembly, and hookup, are major problems.
- Since a majority of the installers weren't technicians or electricians, especially in the early TVRO years, wiring and electronics problems that might be simple for a trained person, could develop into costly and time consuming problems for lesser trained installers.
- Nuisance calls by customers invariably cost at least one or two extra trips per installation, usually requiring two men!
- Because the business appeared to be simpler than it really is, real costs in manpower and equipment often weren't realized soon enough, therefore, price competition was brutal.
- Most TVRO dealers were not diversified, even for rooftop antennas and MATV they were pretty close to being well equipped for any business downturns, even for a few months, were disastrous.

By 1987, the business seemed to have matured somewhat, with makers of poor equipment being weeded out, and dealers and technicians becoming more capable. It is the goal of this book to help satellite dealers and technicians expand and diversify their businesses in order to become stronger and more profitable—to hire better technicians and to do a better job for the consumer. Figure 9-2 shows the principle of a geostationary satellite.

THE SATELLITES

It isn't greatly important for the TVRO technician to know all about the uplink transmitters or the satellite transponders. After all, the site and hardware problems with receive-only equipment is enough of a problem. It is good though to have a working knowledge of what goes on before the signals reach the TVRO dish. Figure 9-3 identifies the major features of a dish.

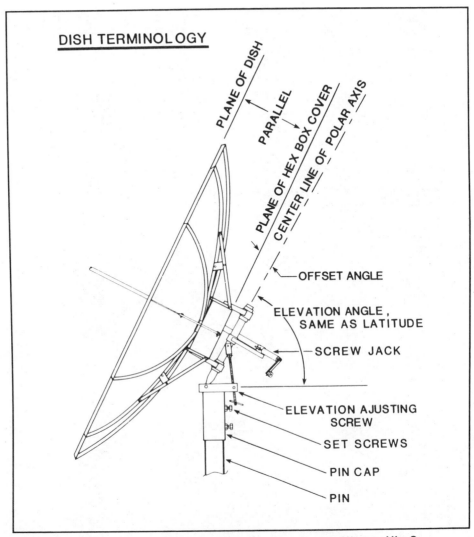

DISH TERMINOLOGY

PLANE OF DISH

PARALLEL

PLANE OF HEX BOX COVER

CENTER LINE OF POLAR AXIS

OFFSET ANGLE

ELEVATION ANGLE,
SAME AS LATITUDE

SCREW JACK

ELEVATION AJUSTING
SCREW

SET SCREWS

PIN CAP

PIN

Fig. 9-3. Drawing showing elevation and declination angles. Courtesy Western Mfg. Co.

Uplink transmission to a satellite is performed by an FCC-licensed broadcaster, using large-sized (10 meters for example) transmitting dish reflectors, and the power of several hundred watts. The frequency band used on U.S. C-band satellites is from 5.925 GHz to 6.425 GHz. (Note the 500 MHz width of the TVRO band.) At the satellite, the solar-powered electronics convert all of the uplink channel frequencies to the common downlink C-band frequencies, ranging from 3.7 GHz to 4.2 GHz. These are retransmitted back to the selected "footprint," or geographic target area of the earth.

The satellites are assigned a location in the Clarke Belt, about 22,300 miles above the equator, orbiting the earth at near 8,000 miles per hour, from West to East. This makes them appear to be stationary, due to the speed at that altitude, equaling the speed of the earth's rotation.

The distance from the earth, 22,300 miles, is important to the installer. If the dish were set to the elevation angle corresponding to the latitude of the location site, it would be aimed in a line parallel to a line between the equator (0 degrees elevation) and the Clarke Belt satellites. A slight declination, or offset angle, toward the south (toward the equator) is needed. This is so that the dish reflector is aimed directly at the satellites. The table in Fig. 9-4 shows the offset angle needed for various elevations around the U.S.. Airports have elevation and azimuth numbers for your area.

Each brand of dish has a different method of pre-setting the declination angle prior to mounting the dish on the pin, or pole. Rarely does this need further adjustment at,

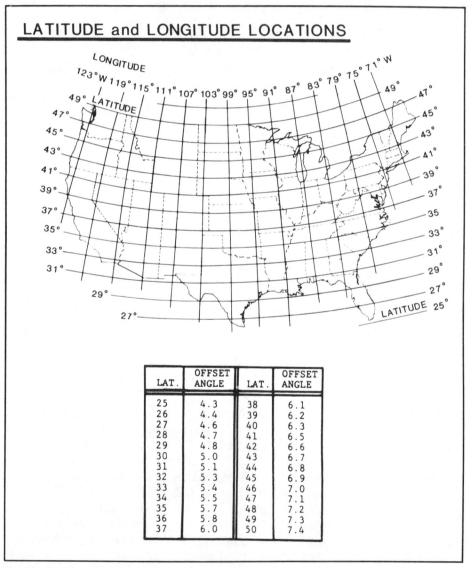

LATITUDE and LONGITUDE LOCATIONS

LAT.	OFFSET ANGLE	LAT.	OFFSET ANGLE
25	4.3	38	6.1
26	4.4	39	6.2
27	4.6	40	6.3
28	4.7	41	6.5
29	4.8	42	6.6
30	5.0	43	6.7
31	5.1	44	6.8
32	5.3	45	6.9
33	5.4	46	7.0
34	5.5	47	7.1
35	5.7	48	7.2
36	5.8	49	7.3
37	6.0	50	7.4

Fig. 9-4. Drawing showing declination offset angles.

or after installation. Without proper initial setting, however, the Clarke Belt will not be tracked correctly. Instead, the dish will either receive the ends of the belt properly but not the middle or vice-versa, according to whether the declination has been set too high, or too low. You might be surprised to find some dish manufacturers suggesting that you add or subtract as much as a degree or two from the listed offset values for your elevation. This is due to the manufacturer's experience with a particular dish, the fitting of the dish elements to the mount, and so forth.

The satellite downlink frequencies (3.7 to 4.2 GHz) are divided into 24 channels. A little math shows that 500 MHz divided by 24 is equal to only 20.83 MHz. That can't be because each satellite C-band channel is allotted 36 MHz for a fully-loaded transponder. Fully-loaded means a transponder utilizing all the allotted channel spectrum. In addition there is a 4 MHz guard-band separating each similarly polarized adjacent channel. (This prevents crosstalk between channels.)

There is room for only twelve 40 MHz channels in the 500 MHz band. By using opposite polarity antenna orientation, and overlapping the frequencies, twice as many channels can be used. Figure 9-5 shows how the vhf channels are blocked in the spectrum.

The video signal transmitted by terrestrial TV broadcasters is an amplitude modulated signal, (AM). Sound is broadcast separately and is frequency modulated (FM). Together the audio and video frequency allotment for a single channel is 6 MHz. To broadcast

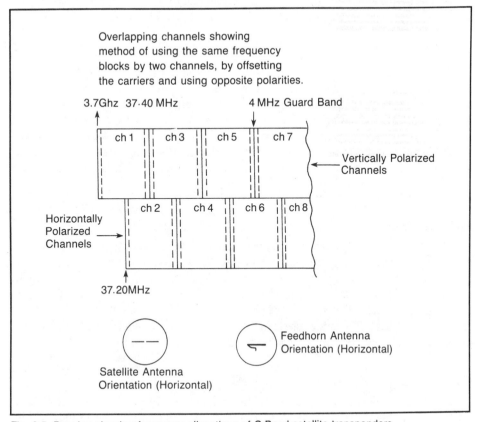

Fig. 9-5. Drawing showing frequency allocations of C-Band satellite transponders.

a satellite signal over such a wide geographic area, with only a 5 to 10 watt transmitter, would be difficult, using AM. Therefore, FM video, as well as audio, is the transmission method.

The use of FM is the reason the 36 MHz bandwidth per channel is needed. Some channels do not utilize the entire 36 MHz band, but use a lesser spectrum, and then devote additional spectrum space to one or more radio broadcasting stations. These radio stations are actually not direct FM broadcasts. They must rely on the TV channel video-carrier, just as the TV video channel audio-carrier must, to be detectable with ordinary TVRO receivers.

RADIO SERVICES

There is a difference between the radio stations that are available as video subcarriers on the satellite transponder channels, and direct FM transmissions from non-video transponders. The non-subcarrier radio services are UPI, AP, Mutual Network, various sports networks, data services, and special transmissions that might be used for as little as one hour per day. They can be state news networks, and individual sports information feeds that provide information about only one team—like the Indianapolis Colts football team, or the Boston Celtics basketball team. WESTAR 3, transponder 2, has these types of services available. Weststar 4, channel 3, is devoted to National Public Radio. There is available space on just one transponder for 400 such radio broadcast services.

The Receivers Differ

The receiver used to select and demodulate these FM broadcasts is not an ordinary satellite receiver; it is more like a business or communications FM radio. These broadcast services cannot be demodulated by home TVRO receivers without the addition of a special demodulator. Microdyne and Harris Corp. are two of the leading companies making equipment for these services. In addition to services used by radio stations, there are a growing number of data services becoming available. CoinNet supplies up-to-the-minute information on the price and availability of rare coins. Stock market data is available through special service networks, as is complete commodity data.

POSITIONING

The satellites are jogged back into position periodically by actuating small hydrazine gas thrusters. The satellites are required to stay within a 70-mile radius parking location in the Clarke Belt. Power for the electronics is supplied by solar cells. Precise aiming of the transmitted beam is required. Some satellites have separate parabolic re-transmitting reflectors that spot Alaska and Hawaii. The Mexican government satellite, called Morales, has a "footprint" directed toward the southern United States and Mexico, making it extremely weak in the northern United States.

Concentrate the Power

Satellite transponders send signals to the targeted land areas using less than 5 watts total power on the SATCOM satellites to around 10 watts on Galaxy 1, 2, and 3. After less than 10 years of use, channel 21 on Satcom F3 quit (July 1986). Other F3 and F4 transponders vary in their power outputs. Channel 1 on both F3 and F4, and channel 6 on F3 are examples of channels that are noticeably weaker. One can expect to have some snow or noise on these channels, even with a 10 foot dish.

To understand how a 5 to 10 watt satellite transmitter can have enough power to be received all over the North American continent, imagine a 4-watt CB radio, using a vertical rod antenna, transmitting in a circular, 360 degree pattern. If you could squeeze the 360 degree pattern into only a 2 degree beam, the 4 watts would then be (at least) 180 times as powerful in the 2 degree direction. It would seem to be a 720 watt CB radio. Then, if you captured the signal with a dish reflector having 78 feet of surface (a 10 foot reflector) it would be further multiplied in strength. Use a high gain, and low noise, rf amplifier to build it up even more. Also transmit an FM signal rather than the more noise-susceptible AM signal. Make the FM signal wide-band, and dither it to emphasize the high frequency portions of the video (the portions that are most suscepti-ble to random noise). De-dither it in the receiver to reduce the high frequencies back to their proper relative levels while reducing any noise pickup an equal amount, and you have a system capable of producing video superior to most off-air broadcasts.

Dish dBs

The actual dish reflectors are rated in dBW. Most 10 foot reflectors claim a 40 dBW gain. That would be doubling the available signal about 7 times, or multiplying the tiny satellite signal by over 100. What is important to the technician installers is that a 37 dBW gain dish is not about as good as a 40 dBW, it is capable of producing only ½ the power. When referring to watts, a 3dB gain is a doubling of power and a 3 dB loss is a halving of the power.

Now we know where the satellites are, and what frequencies they operate on. We know why the small amount of power can produce an outstanding signal on earth, and why we need an offset angle on the TVRO dish. We know the makeup of the frequency band and the type of transmissions we are dealing with. Obviously there is much more to know about the mechanical, and electronic details of the uplinks, the satellite receiver operation, and the satellites. This is an overview, and is information that technicians should be completely familiar with. There are numerous books available on other aspects of satellite technology, and more in-depth information on the topics covered in this chapter. The more technical knowledge you have about TVRO, the better you will be at understanding, and working with the systems.

PROGRAMMING

Once you understand the electronics hardware, you are only half-way home. To install and service TVRO you must become familiar with the available programming. To aim and program TVRO products efficiently, knowledge of the satellite locations relative to your site longitude, and relative to each of the other satellites is necessary. C-Band satellites seldom carry identification of either the satellite, channel number, or name. The only way to become familiar with programming, in order to recognize where the dish is pointed, and exactly which channel you are on, is to spend some time with a sat-ellite dish system.

Some channels that can help you in identification are:

F2—channel 22 AFRTS
F4—channel 13 NESN
T2—channel 10 CBS Net
G3—channel 4 C span
T1—channel 10 ABC Net
W4—channel 13 JISO test pattern
AD—channel 24 CBC bulletin board
S1—channel 21 BTN pattern
W5—channel 2 University Net
G1—channel 12 Nashville network
F1—channel 8 NBC Net

Besides finding the satellites, you must know the programming of each. To tune the channels or center them properly, (especially on single-conversion, 70 MHz systems), the installer must be able to identify whether the channel is correct for the channel position number. Not only is it normally somewhat of an accomplishment to be able to do this efficiently (recognizing over one-hundred different program sources), but scrambling of the premium movie channels, especially on G1 and F3, has made the task for the technician much harder. You cannot normally tell one scrambled channel from another. It is up to the technician installer to make sure all the channels are tuned in properly. You cannot leave a receiver with odd channels appearing on the even numbers, or vice-versa. It isn't the satellite product manufacturers or distributors who receive the nuisance calls to readjust the channels, or set the polarity control for a customer, it is the TVRO technician. There are no short-cuts, and customers in most cases won't even try to figure out difficult receiver operations. The technician must explain in detail how the units work, and caution the owner that it will take two or three weeks to become familiar with the programming, and the nature of TVRO, and to get used to the hardware. After everything has been done correctly, and the customer has been bored with helpful suggestions, and hands-on training in receiver/actuator usage, you will probably average 2 or 3 additional unpaid trips back to the site for any of a hundred reasons, some of these will be explained in Chapter 10.

QUIZ

1. There are fewer than a dozen satellites with viewable programming available to U.S. locations.

   a. ( ) true
   b. ( ) false

2. Each C-Band satellite has 24 channel locations.

   a. ( ) true
   b. ( ) false

3. The C-Band satellites orbit the earth at approximately 8,000 miles above the equator.

   a. ( ) true
   b. ( ) false

4. If the satellite transponders radiated their 5 to 10 watts of power from a 360 degree radiating-antenna, as earth broadcaster do, the received signal would be about 30 percent as strong as it now is.

   a. ( ) true
   b. ( ) false

5. If a technician is familiar with the operation of the hardware brands offered by his company, it isn't important that he have any knowledge of programming content.

   a. ( ) true
   b. ( ) false

6. Dithering (in TVRO communications) is a process used to:

   a. ( ) jog the actuator/dish-positioner exactly onto the desired satellite beam
   b. ( ) reduce the effect of noise on the TVRO video signal
   c. ( ) move the feedhorn antenna rotor to the precise angle
   d. ( ) automatically center the video fine tuning on TVRO channels

7. Setting the dish declination angle is not required if:

   a. ( ) elevation is set properly, using a polar mount
   b. ( ) the satellites were positioned exactly above your site latitude rather than above the equator
   c. ( ) an Azimuth/Elevation dish-mount is not used
   d. ( ) ten foot or larger dish reflectors are used

8. Scrambling or encoding the satellite channels is inconvenient for TVRO dish system owners, but causes no problems for technicians.

   a. ( ) true
   b. ( ) false

9. The individual satellite transponder channels are:

   a. ( ) amplitude modulated and the bandwidth is up to 36 MHz.
   b. ( ) frequency modulated and up to 40 MHz wide
   c. ( ) frequency modulated and up to 36 MHz wide
   d. ( ) amplitude modulated and up to 40 MHz wide

10. 24 TVRO channels can be transmitted in the frequency spectrum only wide enough for 12 channels because . . .

a. (   ) 12 channels are broadcast on AM and the other 12 on FM

b. (   ) most TVRO broadcasts contain only about 20 MHz of information, thus only a small and infrequent amount of cross modulation is ever noticeable

c. (   ) two channels use the same carrier, but different polarity

d. (   ) each channel carrier is offset 20 MHz, and is opposite in polarity to its adjacent channels

Chapter 10

TVRO Dishes, Site Surveys, and Problems

At first, installing a TVRO dish was considered an operation that should be supervised by an engineer with a background in dish construction and design. A few years later, nontechnical people were successfully installing them. In some cases the process is seen as only a little more difficult than putting a Christmas toy together. But any dealer who has more than a handful of installations behind him will agree that TVRO installation can be extremely difficult, that it takes an expert to do it right, and overcoming the myriad of problems that crop up can test the patience and nerves of the most stable person. Figure 10-1 shows a typical 10 foot dish being bolted together.

The first years of TVRO experimentation saw three basic types of dishes. The spherical reflector was constructed on a wood or metal frame, and resembled a segment of a sphere rather than a parabola. Most were large—perhaps 15 feet wide—and 10 feet high, with a feedhorn located 15 or more feet in front. The idea was to try to locate one satellite. Later, the feedhorn was made adjustable, so that one or two adjacent satellites could be picked up without rebuilding or reorienting the basically permanent reflector. Figure 10-2 illustrates the important parts of a satellite dish.

AX-ELs

Parabolic dish reflectors were easier to produce and aim at multiple satellites and thus became the standard. Before polar mounts were available, AZ-EL (Azimuth-Elevation) mounts were used. The dish could be repositioned East or West by turning the entire mount. Elevation is changed with a separate adjustment that tilts the dish up or down. Some AZ-EL mounts have been made with two positioning motors to do

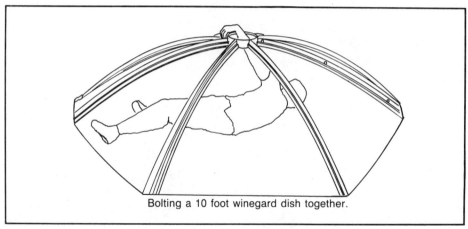

Bolting a 10 foot winegard dish together.

Fig. 10-1. Photo of dish assembly process.

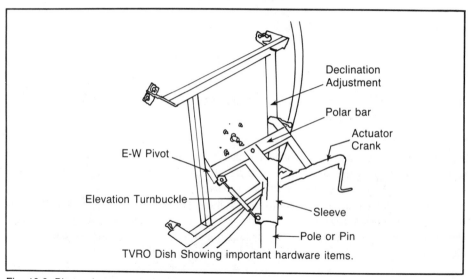

TVRO Dish Showing important hardware items.

Fig. 10-2. Photo of cassegrain feed type satellite dish.

the same job remotely. AZ-ELs are able to zero in on a satellite exactly. Polar mounted dishes can track the satellites precisely, but there is likely to be some compromise that, while satisfactory, might not be optimum for all satellites. Some attempts have been made to make AZ-EL mounts with two programmable controllers to automatically lock in each satellite; however, the polar mounted dishes are so satisfactory, that the two-motor idea is finding tough competition. You can find an example of an AZ-EL mounted dish at most cable company sites.

USCI Dishes

Another type of reflector using an AZ-EL mount was produced by Prodelin for USCI (United Satellite Communications, Inc.) in the early 1980s. It was a small one-piece, fiberglass reflector that was neither parabolic, or spherical. It was designed to operate on one polarity only, and to act like a much larger dish. It operated on the signals from a satellite producing about 25 watts of power, rather than five watts, and thus had an excellent picture.

USCI's scheme was to sell programming on a monthly fee to rural cableless residents. Five channels were offered for a time in a Midwest experimental area. USCI marketed the dish for many months before the larger, polar-mounted dishes became the *de facto* standard way of doing things. Five channels for $25 per month would have been considered an exceptional value ten years ago. At the time USCI made its move, the industry was headed toward a hundred or more channels, and no monthly fee. Thus USCI went out of business.

The USCI dish was easily mounted on a roof, since the mount (using a 2.5 inch i.d. pole), was light, and had little wind loading. A disadvantage was that the AZ-EL mount would allow reception from only one satellite.

A Wide Arc

From time to time, hand cranks were used on parabolic polar-mounted dishes. Several hundred dollars could be saved by not investing in an electric motor-driven positioner. An additional incentive was, from 1980 to 1985 almost all of the premium programming was on three satellites, located within 12 degrees of each other. With a couple dozen top-flight channels available, who needed a $500 automatic positioner? From time-to-time, events took place on the Canadian AD, or on RCA's SATCOM 4 satellite, that became desirable. New satellites took their positions, and the desire for an ability to track all satellites easily, caused the polar-mounted dish with automatic positioner to become favorable. The polar-mounted dish is mounted on a bar, aimed directly at the North Star, the bar swivels East or West. If the dish is slightly tilted downward, to where its boresight beam intersects the satellites 22,300 miles above the equator, the dish will track all of the visible satellites with only one motor, or adjustment movement.

To install an earth-mounted pole and dish properly, you need the ability to dig a trench into the house, to bury the wires. You need more muscle to dig a 3, or 4 foot hole, to mix and pour cement around a heavy 3- to 4-inch diameter steel pole 8 or 9 feet long. More muscles are needed to hoist an 8, or 10 foot diameter dish, and its mount onto the pole. The hard part being done, all that is needed, is to adjust the angles, and behold, Satellite reception!

There are hundreds of cases where the description of the installation process, pretty-well covers the job. How then, can TVRO installation procedures take up two chapters in this book? The answer is most installations are not that simple. Here are some of the reasons:

- Since 1985, the most popular dishes have been black mesh. These dishes require from two to 20 man-hours each, just to put the pieces together.
- Many installations are sold with an agreement that the dish will be mounted well off the ground. If the dish elements and adjustments are higher than you can reach, time will be needed for ladders or scaffolds. A crane might be required to raise the dish onto the pole.

- Entry to a building can require boring under walks, fences, pipes, wires, drains, septic fields, and trees. Each of these problems consumes time.
- Each new dish brand requires a virtual apprenticeship to learn its correct assembly. Poor instructions are the rule, and wrong assembly is easy.

Some dishes won't work, even if put together correctly. They all work, but whether the dish works as well as it should is frequently a question that isn't answered completely, until the electronics have all been substituted, and each adjustment has been readjusted to prove the fault is not anything other than the dish.

Dish mounts frequently do not fit the pin tightly, the play causes a slight tilt from perpendicular that can reduce reception quality and complicate aiming.

Running wires 100 to 300 feet is pretty straightforward, except that apprentices and helpers will invariably put a staple directly through the coax, or actuator cable. Feeling bad about their lack of ability, they hope the staple causes no harm. You will spend an hour or so troubleshooting the problem.

Dish owners are affluent. Most have nice homes. Most want wires run in attics and crawl spaces, snaked through walls and false ceilings, and put neatly through a hole you will drill in a 12 inch thick foundation. These things will require much extra time, and ingenuity.

LNA (Low Noise Amplifier) weather covers and actuator boots will require an extra half hour or longer to install properly. Many LNA covers won't fit the electronics, and a time-consuming decision to remount the electronics on the LNA support, or to go back to the shop and get a larger cover, will result. All changes in plans take time.

The customer will expect several TVs to be hooked up and your salesman will not have explained the extra costs. Nor will it be included in the bill. You will explain it and charge for it, or you will pay for it yourself. Each extra wire run takes time and materials.

VCRs are standard equipment. You are expected to hook up any and all brands, about 50 percent will require some form of channel tuning, or switch resetting. Connecting a VCR as the set-owner would like, can be impossible, or will require a video switchbox. Whatever you do, it will take more time.

It is not unusual to find one or more pieces of the TVRO system inoperative or defective when delivered. These problems can range from a dead LED signal-strength indicator to an incorrectly-sized actuator, or a missing waveguide for the feedhorn.

Sometimes the actuator damages the antenna the first time it operates. You try to bend the metal frame back into shape. Your helper pokes a hole in a mesh dish panel. The pole diameter is wrong for your new style dish. The receiver won't tune the lower two channels. That's only part of the problems that can and do occur. Some are preventable, others are not. Figure 10-3 shows a dish panel damaged while demounting.

Simple things—your receiver, LNB (Low Noise Block dc), or switchbox—must have RG-6 cable, rather than RG-59. The RG-6 center wire is #18 instead of #22, as in RG-59. The RG-6 won't fit the female sockets. How can that be? Couldn't the female sockets all be spring tensioned slits that accept either size wire? They could be, but they aren't. You will put a barrel splice on your RG-6, then a male-to-male connector into the small female receptacle. Simple enough. The problem is that you might be out of splices, or adapters, and you might be 250 feet away from your truck where some spares are. Chances are the problem will cost you only a few minutes.

Fig. 10-3. Dish damage during deinstallation.

INSTALLATION PRACTICES

When experienced installers are asked "How do you overcome the problems with installations of dishes and antennas?", the answer we receive most often is: "Take the time to make a complete site survey."

It is easy to take a quick look at the property, ascertain that trees aren't a problem, and quote the job. You've probably done several installations, and pretty well know what has to be done. In the hustle and bustle of the satellite/antenna business, time is at a premium. Why waste time with details?

It isn't a waste of time to do a complete survey. Start with the dish location. Look at all the possibilities. Most dishes are best mounted in the back yard. But if the back yard has several underground electric lines, sidewalks, or no house entry, the side yard would be better. A pole mount from ground to roof level can sometimes be a better answer. I've seen salesmen select a location that had no advantage at all over one 80 feet closer. Selecting the wrong entry method can be a problem. Some crawl spaces are too small, too wet, or too infested to use. Better to go around the house, and go straight through a wall. Sometimes drilling down through two or more floors is a practical cable run path. If the closets are all packed with junk you will have to remove, and replace; and the house is built of hardwood that is difficult to drill through, and harder to fish wire through, you might spend some frustrating time doing the job. Concrete blocks aren't difficult to drill through with a masonry drill, or star chisel, but poured foundations, and 12 inch block walls are real challenges. Customers will suggest the crawl space is accessible, but you or your helper will have to work in it. If it's a problem, don't do

SITE SURVEY WORKSHEET

DATE _____

CUSTOMER NAME_____ ADDRESS_____
CITY_____ ZIP_____ PHONE (___)_____WK _____
DIRECTIONS TO HOME _____
_____

SPECIAL PROBLEMS OR INSTALLATION SUGGESTIONS _____
_____

ENTRY METHOD ( ) THROUGH WINDOW, ( ) FISH THROUGH WALL
( ) THROUGH BASEMENT, ( ) THROUGH CRAWL, ( ) THROUGH ATTIC
( ) OTHER _____

MOUNT: ( ) 3' POLE, 8' POLE ( ) LARGER POLE _____FT?
         ( ) LONG POLE ASIDE STRUCTURE, ( ) ROOF MOUNT
         ( ) ON PATIO OR WALKWAY MASONRY, ( ) OTHER_____

CUSTOMER PARTICIPATION:
                    ( ) CEMENT WORK, POLE, TRENCH, ENTRY HOLE
                    ( ) NONE
                    ( ) EVERYTHING BUT AIMING

SKETCH OF YARD

EXTRA SERVICE ( ) VCR HOOKUP
              ( ) VCR RUNING, U/V CHANNEL SELECT
              ( ) ADDED TV # _____   ( ) SWITCHBOX
              ( ) WIRE TO SEPARATE STRUCTURE
              ( ) U/V ANTENNA WORK REQUIRED? ( ) ROTOR?
              ( ) PUT IN SUBSCRIPTION TO PROGRAM GUIDE?
              NAME OF GUIDE AND PAYMENT _____
                   $ _____

WILL CUSTOMER BE HOME? ( )YES, ( ) NO ( ) DOOR OPEN ( ) KEY?
ANY RESTRICTIONS? ( ) YES, ( ) NO

OTHER INFORMATION

Fig. 10-4. Site survey worksheet.

it; or, charge what it is worth to make the extra effort, and spend the time it takes to do the job right.

A worksheet is handy for the site survey. Take it with you, along with your compass, and declinometer. Take the equipment invoice or bill, and check the site carefully. Figure 10-4 is an example of the worksheet.

Before heading out for the site survey, call the customer and make sure someone will be home. Reaffirm the location, or you will spend an hour or so finding the house. The time will not be paid for by your boss, or the salesman. Neither took the time to get the directions explicitly, after all, why should they? They don't have to find it, you do. Call the customer, confirm the directions. Check your county plot maps. Question the customer about any unclear directions, and repeat them. If it says the first house after the tee, make sure you know they mean the first house on the *right*. After you are assured it is the first house on the right, make sure the customer says "Yes, it's on the west side of the road." Frequently, people don't know the name of the road or street they live on! Neither your salesman nor the customer pay for *your* time sifting through the clues to get you to their house.

At the Site

At the site make a drawing of the lot, and the location you have chosen for the pole. Draw a trench line. Mark each potential problem, and decide how to overcome them. (Example: sidewalk is two feet wide and is expected to be four inches thick. Take a masonry drill bit and loosen the gravel or fill-dirt under the walk. Use a small garden hand-spade, or a length of antenna pipe to clean out the tunnel.)

Show the entry location into the house, and how you expect the wires to run. Indicate if PVC pipe is to be used at the house to protect the wires from gardeners, dogs, lawnmowers, and so forth. If the crawl space is small, devise a plan you think will work to do the job easiest, cleanest, and best. If the carpet is to be drilled, let the customer know exactly where and how big the hole will be. If wires are to travel any distance down the toe molding, are they to be stapled neatly, or just lay on the carpet or floor? Is a VCR involved? Is more than one TV set to be hooked up? Is an extra charge included in the bill for any extra runs and hookups? Will you have to change any outside antenna wires? Is a stereo to be hooked up?

Everything must be noted in detail. Extra work, extra wire, and extra fittings should be included in the charges. Learn to say no to customers that feel you are in no position to deny their extra requests: "No, you can't make their 20 year old TV set look better." "No, you can't hook up their grandmother's set next door for free." "No, you can't replace the chimney mount on their antenna while you are there." "No you can't fish the wires down through a blocked wall space and have a wall-plate plug installed for all of the dish wiring, at no extra charge."

If the site survey is completed to your satisfaction, the next step is to check to make sure all of the correct hardware is available and working. Again, check the sales invoice, and physically locate the equipment. Tag it with your customer's name, and get the serial numbers on the invoice or site worksheet. Is the pole the right diameter and length? Is a house bracket, or roofmount needed and available? Do you have plans for the cement? Do you have a post-hole digger, pick, spade, hoe, and other tools? Are your walkie-talkie's working, and a spare set of batteries handy? Do you have sufficient wire of the proper size? Will you need a hoist? Do you have a copy of the latest program guide? You should go over everything in detail, as one extra trip back to the shop to pick up a piece of PVC, the compass, the drill, or a video switchbox will cost you dearly.

The Dirty Work

Planting a satellite dish pole is done skimpily by some, and elaborately by others. I have see installers use as little as four bags of Sacrete® around the pole. Their hole diameter is less than a foot. Others use a full yard (3 × 3 × 3 foot) of concrete. Both are probably extremes for ordinary ground-mounted eight-foot or ten-foot dishes. Four to five times the pole diameter is a good rule in clay, more for sandy, or loose soil.

You can use a powered two-man post-hole digger, a back hoe, or shovels, and a hand post-hole digger. Put a bag or two of cement in the hole, then set the pole in. Don't bother trying to level it until you have four or five bags of Sacrete® in. Be sure the pole has an anti-twist rod, or angle iron, or simply a large bolt in the base. After the pole is fairly well-supported by the cement, install conduit or PVC up the side, and tape it. Make a gentle curve into the trench. Make sure the conduit or PVC is not above the mount pin-sleeve near the top of the pole. Cement the PVC in place, and make the final level checks to see that the pin is as close to perpendicular as you can get it. Go all around the pole with your level. The top one foot is what has to be level. If there are bumps and globs on the pole surface, file them off. If the pole is too small for the dish sleeve, be prepared to insert some shims in it.

QUIZ

1. The majority of today's home TVRO dishes are:

   a. (   ) parabolic in shape
   b. (   ) spherical in shape

2. A parabolic reflector that uses a second, much smaller reflector to channel the satellite frequencies down a tube or waveguide to the rear-center of the dish, is called a Cassegrain feed dish. This secondary reflector, like the more common feedhorns, is located . . .

   a. (   ) at the sidelobe
   b. (   ) at the prime focus point

3. The polar mount has a rectangular, or cylindrical shaped subsection called a polar bar. In any location in the northern hemisphere, on a properly adjusted mount, this polar bar has its North end pointed at . . .

   a. (   ) the big dipper
   b. (   ) the North Star
   c. (   ) the North Pole

4. If a polar-mounted dish is set to the correct elevation of the site, and the correct declination is zero degrees, the location is:

   a. ( ) at the North Pole
   b. ( ) at the South Pole
   c. ( ) at the equator

5. Because of universal fabrication methods, and good instructions, most "kit" type mesh satellite dishes can usually be assembled by two men in less than . . .

   a. ( ) 2 man-hours
   b. ( ) 10 man-hours

6. Given today's zero-defects manufacturing techniques and tight quality control, it is unusual to have any component of a C-Band satellite system fail at installation.

   a. ( ) true
   b. ( ) false

7. A site survey is usually required only if:

   a. ( ) tree branches may be a problem
   b. ( ) the salesperson or technician has little experience in the business
   c. ( ) maximum potential problems are to be uncovered in advance

8. The diameter of the cement hole for a four-inch i.d. pole eight-feet high should be:

   a. ( ) 6 to 8 inches
   b. ( ) 16 or more inches
   c. ( ) 3 feet by 3 feet

9. A/An _____ should be attached to the pole 6 to 12 inches from the bottom in the cement hole

   a. ( ) piece of PVC conduit
   b. ( ) anti-twist rod
   c. ( ) ground rod

10. The major concern regarding perpendicularity of the pin is:

    a. ( ) that the top of the pin is cut evenly
    b. ( ) that the top 12 inches of the pole is perpendicular
    c. ( ) that the pole is perpendicular at the ground point

Chapter 11
TVRO Installation: Good Practices

While the cement is setting up, dig the trench. It's nice to have an electric or gas-powered trencher. If you don't have one, dig. A shovel will work. Turn the sod over, so that you can flip it back in place after the cable wires are laid six to eight inches deep. A neater trench can be dug with a spade. Even better is a hand trencher made from half of a 16 inch serrated farm disc. This trencher won't work well in hard dry clay, in rocks, or if the ground is frozen more than an inch, but it can dig a 100 foot trench in less than an hour, usually. After splitting the soil with it, you wiggle it back and forth, hollowing out the bottom of the six to eight inch trench. When you are through digging, lay the wire in the slit and then poke it down to the bottom with a broom handle, or large wrench handle. As you walk back and forth from the dish to the house to adjust the dish, to change inoperative parts, and to do other chores, walk on the trench. By the time you are through, the trench can be unnoticeable. Sometimes dirt falls back into the trench in a spot or two. The wires usually can't be forced through the dirt, it has to be scooped out by hand, or the trencher used again in those problem spots. Young helpers will not take the time to clean out any shallow places, so you will have to go with them on the first cold, miserable, windy day, when the customer calls up and complains about the wires pushing up out of the ground. After redigging a portion of the trench, and perhaps splicing the wires, or replacing the entire cable (at your expense) the helper will usually do better the next time. The economics of buying a power trencher for $1000 or so are: An apprentice costs you $20 per hour minimum, the trencher, by saving an average of an hour on each job, pays for itself in 50 jobs, plus, you then own it!

Cement walks are a hindrance. Like most problems, they seem worse than they really are. A larger problem is a walk or patio with no break between the house. You can sometimes find a severe aging, or weather crack, that can be widened enough for the wires, then recemented, leaving the customer with a neater area than they had before you arrived. Most of the time, you can find a longer path that can avoid the walks. It's easier to dig an extra 50 feet of trench, than burrow under a well built 4 foot walk.

Some installers put the entire cable or wire set in PVC conduit from pole to house. Most don't since the wires are already well insulated against moisture. Most installers do use PVC up the pole and a foot or two into the trench, making sure the PVC is not cemented in the pole hole with a pinch in it. At the house entry, use PVC or other protection, so that plants, animals, lawnmowers, or trimmers won't damage the cable. If you can drill an entry underground into a basement or crawlspace, then you need no PVC, but that is a pretty good job in itself, getting room to drill a hole that deep under the ground surface.

SPLICES

At all costs, try to keep from splicing any of the wires, except at the pole where they can be waterproofed, or inside the house. If you do have a repair, or mandatory splice, you can use a telephone company splice-package, or some other commercial method. I twist each wire individually, use barrel splices on the coax, and use electrical tape on each wire individually. Next tape the individual wires together in a bundle, use silicon sealant on the entire splice. Finally encase the splice in a piece of PVC, tape it after putting sealant in the lengthwise cut needed to get the PVC over the wires, and fill the ends of the PVC tube with sealant.

If worse comes to worse, you may decide that mounting the wires overhead makes more sense. In the desert you will encase coax in PVC to protect it from severe heat. In the far North you will need strain relief to protect against ice storms and howling winds.

THE ENTRY

Drilling holes in someone else's home is scary. You could drill through an electric line, water pipe, the phone junction box, meet up with a steel plate, or simply hit an extra hard stud. You can punch out a nasty hole in paneling, siding, toe mold, or a basement wall. No amount of words can keep you from being a klutz! The best advice is to go slowly, take time to survey the proposed entry site, and try to anticipate any possible problems.

Customers won't tell you about unusual or potential problems. After all, they don't want any extra charges. They do expect you to fix up anything you damage or deface.

You will need a star drill or two, at least 14 inches long. You will need some masonry bits (at $30 each!). A ⅜-inch wood drill bit can be used for most uhf/vhf antenna holes. Making two or three holes in a row for your cables can be neater than drilling one hole an inch in diameter for them. Coathangers, a piece of aluminum grounding wire, or guy wire are good to use for wire pulling. Tape only one of the cables to the pull wire until you get it through, then tape the second wire on it a yard or so back, then the last one. Be careful when you have just-enough-room for the wires. If one gets crossed over the bundle, you can get a jam that won't go either way, causing a time wasting delay.

Once inside the home, neatness is the rule. TVRO wiring is ugly and bulky. Look for ways to hide it. If you do have to cross highly visible areas, staple the wires (carefully) in neat rows along the baseboard, or around the molding. Often you can improve the existing uhf/vhf antenna/rotor wire runs at the same time, making a friend out of the customer.

PUTTING THE DISH ON THE POLE PIN

Four or five competent people are required to place most dishes on the pole pin safely. Most dishes, however, are precariously put on the pole by two or three lucky men. Ordinarily, helpers will at first attempt to get the dish aimed 90 degrees straight up! This is with the idea of then letting the sleeve down on the pole, which also is aimed straight up. This usually won't work because the mount sleeve will be at an angle—usually 45 degrees or less. After fighting the pole pin and each other, the dish is usually broken or ripped and the helpers blame you for not letting them do it their way.

Most experienced installers will roll the dish to the pole, put the mount back against the pole, with the dish on edge. They will make sure the sleeve is swiveled to a position as near parallel to the dish face as possible. With one man assisting and guiding at the pin, and the mount back, the other two lift from the edges, vertically, slipping the dish up and carefully sliding the sleeve over the pin and then down into place.

With all the grunting and heavy lifting, there may be a better way. One way is to install only aluminum mesh or perforated dishes, preferably no more than eight or nine feet in diameter. If the dish is parabolic it will perform better than 90 percent of the 10 foot or 12 foot dishes on the market. These large antennas can be lifted onto the pole using a one-man crane or a pole hoist (see Fig. 11-1). L.S. Engineering, 1110 W. Crocker, Avoca, IA 51521, makes an aluminum hoist that solves the weight problem.

ROOF MOUNTS

Roof mounts are difficult in many ways, but not impossible. Charge for the extra danger, pole length, and problems. Sometimes the dish can be built on the pole, or the roof. Others will require a crane, and some roof monkeys! Most proposed roof mounts can be changed to ground mounts on long poles attached to the roof. The pole can be lag-bolted into concrete at the base, if it is put on a patio or driveway of sufficient thickness. It can be cemented in like any other pole, and bracketed to the structure near the roof line. Brackets are available commercially, or your local welder can make custom adjustable brackets for any situation. If the bracket is not lag-bolted into solid studs you can install a mounting board that is bolted to solid studs. If you put the collar into lesser materials (older wood or mobile home frames, or awning framing), realize that the base cement is going to do the real work of keeping the pole from swaying. You might want to use a bracket with some spring in it, and/or some shock absorbing material around the pole, before the bracket clamp is tightened.

The idea is to give the pole some support at the collar, but to keep the potential ¼, or ½ inch of sway, from pulling against the structure. At the same time, you have some additional bracing to aid the concrete in the base.

A 20 foot pole, by itself, is heavy. Using one man on the roof and two or three on the ground is needed to position it. You can use a sign crane if one is available. A 20

Fig. 11-1. Photo showing how a dish crane is mounted on pole.

foot pole should be cemented into the ground deeper than an 8 or 9 foot pole. Three feet deep is enough for a good mount, four feet for a 10 foot pole, five feet for a 15 foot pole, and six feet for a 20 foot pole. Fill the *pole* with cement for several feet above the ground level to further strengthen it, and to spread the pivot or break-off point.

For actual roof-mounted dishes, there aren't any secrets. Several firms make roof mount brackets. Your local welder can make them out of angle iron. The best plan is to figure on drilling through the roof and putting ½ inch threaded stock, or long bolts, through the roof sheeting, then through two 2 × 6 inch boards mounted under at least three rafters. You should use nothing larger than an eight-foot perforated dish. If the construction permits it, put two or four 4 × 4 inch studs upright, under the roof, and over the ceiling joists to prevent dish weight from warping the roof later. On non-trussed roofs, such as in vaulted ceiling homes or log cabins, if the peak of the roof is merely toe-nailed together, add some ¼ inch steel, bolted tie-togethers, at the peak, to discourage any loosening of the roof elements. Before you start, get plenty of roof patch and be sure to seal the bolt holes before, and after you put the bolts in. If you are placing the roof mount on boards, overlaying shingles, seal the entire board, double, from end to end, under, and at the drill points.

I don't discourage roof mounts. There is a place for them and some of my most satisfactory installations are roof mounts, or above-roof installations. They require double or triple the effort in most cases, and it has cost a couple days extra time learning this. Remember that everything you do requires trips up and down ladders getting to the feedhorn or actuator. In cold weather or snow, or extremely hot locations, you are going to have to go back up and service it. In the beginning you will do this for free. Be sure to include the possibility in your job price estimate.

AIMING

The first step in proper aiming, is to have a perpendicular pole in the ground. Secondly, you need a good dish. Most dishes are not good. Some spun-aluminum dishes have ripples on the surface. The first mesh types were too heavy, and too weak to support their own weight. Many dishes weren't designed to be perfectly parabolic. Some eight foot dishes have three inches around the outer rim that is folded down 40-60 degrees, in order to strengthen the rim. This makes the size of the actual parabolic area only 7.5 feet. All kit dishes should be put together on a flat surface. Most are constructed on the grass, on a deck, on a driveway, or at best, in a garage. Chances that the dish is perfect are slim. The early four-section fiberglass types were made in a mold so that it was difficult to bend the individual sections. The only mistakes a dish assembler could make would be to bolt the sections together unevenly or to put the mount frame on in a manner that warped the dish. Since many fiberglass dish sections wouldn't fit evenly, and the mounts weren't designed to bolt to the pre-drilled panel-bolt holes easily, a lot of fiberglass dishes are operating in a warped condition. The reason a Bentley 8 or Winegard 8 or 10 work, is because each of the segments is pre-formed and permanently attached to rigid ¼ inch aluminum framing. The parabolic shape is maintained.

Most UPS-shipable mesh dishes have separate ribs that are inserted into a center hub that is rarely over 16 inches in diameter, even on a 10 foot reflector. Following the instructions explicitly, building on a flat surface, and carefully transporting it, you might get a workable dish.

Signs of a Bad Dish

Signs of a good dish are: you sight across the edge of the dish and see the edge closest to you as a straight line. The opposite side should also appear as a straight line, parallel with the closest side. If you see a slight bulge up or down on any portion of the surface, you have a less than perfect dish. It might work satisfactorily enough to get you by, but it will be a 10 foot dish performing like an 8 foot one should. Trying to put the dish back on a flat surface to straighten it out is often a lesson in futility, as is loosening bolts, guying, and using turnbuckles to attempt to eradicate the bulge.

Another sign of a bad dish is that the focus seems off. When you move the satellite through the satellite belt, as you come close to a satellite, you will see a faint image of the picture. Rather than getting clearer as you get closer, the image will fade out, then come in strong as you reach boresight. As you begin to go past it, the image will again fade out, but then get a little stronger before you exit the beam entirely. These strong sidelobes aren't always a sign that the dish is not satisfactory, but on a dish that is not performing correctly, the sidelobes are an indication that the surface is not parabolic.

FOCUSING

Before aiming the pole-mounted dish, you should set the feedhorn opening to the exact distance from the faceplate at the center of the dish, as recommended by the manufacturer. Unless it is stated by the manufacturer, it is not easy to guess the actual diameter of the dish, the depth, or whether the dish was designed to be *exactly* parabolic. It is usually difficult to mathematically determine the proper focus point using the formula:

$$F = \frac{Diameter^2}{16 \times Depth}$$ (Using this formula, be sure to use inches throughout, not feet.)

Whatever you do, don't adjust the focus for best picture until every other adjustment has been peaked. If you get the focus point off a little, you can really mess adjustments up by trying to compensate for wrong focus by adjusting elevation and declination again, or trying to reset the East-West position a little better. You can be trying to zero in on a lobe, and can spend hours fiddling with the dish, deciding if the dish is bad! If you lose the instructions for building the dish, take the time to call the distributor or manufacturer, and get the exact focal point distance. When you do decide to peak the focus visually, be sure first to mark the correct distance, so you can always return to it easily.

HOOKUP

So the dish is a good one, you've installed others like it successfully, the pin is perpendicular, the focus is set to the manufacturer's specifications, now what? Its time to hook up the wires. Feed the LNA/dc and rotor wires up the feedhorn tube. You'll need F-59 adapters and barrel splices to get the RG-6 wire to push into the F-59 sockets on the LNA/LNB, so plan on it. If you force the RG-6, you will push the center wire back in the coax. Your connection will be by touch only, not a secure socket-to-plug one. It will work just fine, until tomorrow! Do it right. If you're using a separate downconverter for a 70 MHz system, you will want to mount it alongside the LNA for convenience and neatness. If you use N-type connector splices and elbows, be extremely careful. The N-connectors are notorious for going bad. They merely open up and have you changing components before you realize the connector is the culprit. On block LNB systems, you have none of these problems, a single RG-6 is the only connector needed.

Antenna Polarity

Polarotor, Unirotor, Omnirotor, and other brand names of antenna polarity positioners are of three types:

- No feedhorn probe to position at all. The signal reflected off the dish into the feedhorn opening is channeled directly to the fixed LNA/LNB pickup probe.
- A dc motor drives a movable antenna probe in a 360 degree circle.
- A servo motor, with 180 degrees of travel, uses dc power to operate a chip inside the horn housing unit, using square-wave positioning pulses from the receiver.

DC motors are easy to understand. If you can hook up two wires and plug in a 9-volt battery, this type of antenna positioner works. Few are used lately, because they are slow, the battery must be changed annually, and the energizing switch must be held down until the proper position is obtained. They do work, and they are very reliable.

Dual block systems don't use a positioning motor for the antenna probe because they don't use an antenna probe inside the feedhorn. The horizontal and vertical signals from the satellites are reflected into the feed opening. The two polarities are directed or channeled into their respective polarity openings, and then to the LNA, or LNB pickup element at the LNA/LNB entrances. Horizontal and vertical polarities are separated entirely due to the waveguide locations, and dimensions. Microwaves will travel down a rectangular pipe or tube. A rectangular pipe will reject signals perpendicular to the short dimension, but allow a signal of the same frequency through the wider non-perpendicular dimension. By constructing waveguide openings mounted perpendicular to each other on the horn, the odd and the even channels are separated, and directed to their respective LNAs. The receiver switches either of the two LNAs to select odd or even polarity channels.

Because the rotor motors require three wires generally, some feed makers supply LNSBs—Low Noise Switching Block feeds. These use the same waveguide method of separating even and odd polarities, directing the signals into separate LNAs built into a single LNSB horn. A single control wire is used to switch either of the LNAs, the unit is compact, and the mechanical rotor movement is eliminated. The return wire for the switch is the coax shield used for the 950 to 1450 GHz signal to the receiver.

The servo motor was the most widely used until 1987. It can be switched 90 degrees almost instantly. It works a long time with no problems if properly installed, and it uses little power. To install a servo motor style horn properly, three wires must be hooked up correctly: ground, five volts, and pulse. Because of the long wire runs in TVRO, several ohms of resistance in the wire, even if hooked up backwards, cause little damage. Rarely does a fuse blow, usually the rotor just doesn't move. Switching the wires to the correct position solves the problem. At the dish, the three wires can be connected to the rotor cable using small wire nuts, electrician's snap connectors, or other splices. Since it seems that you often have to substitute something at the horn, the splices frequently must be undone. Simply stripping, twisting, and taping is a common practice. Untrained helpers will rarely splice the wires correctly. Even if they do get the wires twisted tight, and each individually taped, at some time a single strand of wire will poke a hole through two layers of tape and cause a short. Be careful! Figure 11-2 shows a typical feedhorn servo motor.

If you position the feedhorn so the the LNA (mounted on the rear of the feedhorn) is parallel to the earth, you will probably find one or the other polarity will be in its optimum signal position when it is positioned all the way to one end. In this position, rotors will continue to attempt to move the antenna probe slightly further; working too hard, and eventually burning out. To eliminate this, you have to get in the habit of positioning the feedhorn at a 45-degree angle to the ground and the polar bar the dish in mounted upon. This way the 90-degree travel of the servo motor and antenna probe, will fall about half-way between its 180-degree maximum extremes. It is tempting to leave an incorrectly oriented feedhorn as is, but it is worth your time to go back out to the dish and orient it, to avoid a warranty repair call to replace the horn in a week or so. Be sure when

Fig. 11-2. Feedhorn servo motor.

readjusting the horn, that you don't change the focal distance, even a hair. Make sure you have installed a weather cover, and that no holes are in it big enough for birds or wasps. That means tight. If you leave wide gaps in the cover you will have bird nests to contend with, and wasps to fight.

ACTUATORS

Few handcrank positioners are used today; however, they mount just like motor drives in many cases. A saddle-clamp attaches the outer stationary shaft to the dish mount bracket. The telescoping, inner-shaft and eyebolt attaches to the dish brace, a couple feet out from the dish-back center, and hub. Two things are important: Make sure the saddle-clamp has a nylon washer between it and the dish mount actuator bracket, so that the two can slide against each other easily, and freezing rain won't bind up the dish. Make sure you use a long enough bolt, and the proper shim sleeves on it at the eyebolt. It might look fine while you are positioned to the West, on F1 or G1, but when a linear actuator moves toward the east (or west on the West Coast) the two-inch diameter actuator shaft end may not have clearance against its brace. It will bend the brace, or, if merely a tight squeeze, it can bind up just enough to blow fuses at the extreme travel. Figure 11-3 shows a typical actuator motor.

CONNECTIONS

Three types of linear, or horizon-to-horizon actuator sensors are common:

- Reed switch
- Optical sensors
- 10 kilohm resistor pots

The 10 k resistor type used by Janiel, Drake, and others in the past, are nice because a helper can hook up the two 36-volt motor power wires across the pot and no damage

Fig. 11-3. Actuator motor housing and gear box.

will be done. Do this to a reed switch, and it is burned out. The optical type (Uniden) has a transistor and capacitor in it that blow apart when the 36-volts is applied to the sensor, instead of the motor. Teach your helper to locate the larger motor wires in the actuator base, and to have great respect for them.

Obviously, the three-wire 10 k pot-type sensor (five wires total) must be hooked up properly to work. Start out using a color code sequence that you can remember for all your wire hookups, like the values for resistor color-code.

Color	Value
Black	0
Brown	1
Red	2
Orange	3
Yellow	4
Green	5
Blue	6
Violet	7
Grey	8
White	9

If you get in this habit, you will not always be wondering what sequence you hooked up the sensors, at the now sealed actuator, or any other multi-wire terminal strip.

The two-sensor-wire reed switch type (M/A-COM, MESA, MTI) merely opens or closes a magnetic switch relay position, and can be hooked up with no regard to which of the two sensor wires is on terminal 1 or 2.

Most dishes can be made to track F1 (or F5) in the West, to S2 in the East, using an 18 inch actuator shaft. While it might seem that a 24 inch one would be better, allowing some room for error, you want to use the smaller size where possible. That is because greater resolution, and greater accuracy in programming and positioning occurs when you use the entire length of the shaft, and thus the maximum turns of the sensor counting unit. Think about it. If you have an actuator that uses only ten percent of its travel, and it is using a 10 k pot-type sensor, you are trying to sort out over 20 satellite locations, using only $1/10$th of the pot travel, and only 1000 ohms. This is less than 50 ohms per satellite location. If you were using the entire shaft length, the pot would be changing over its entire 10,000 ohm travel and would have an average of about 500 ohms per location to lock on. With the reed switch, it might be the difference between 40 to 50 switch clicks per satellite, or only 10 to 15. Ten or 15 clicks might be less than the mechanical play in your dish mechanism.

Rain

The worst enemy of an actuator is rain. Most actuators have a weep hole in the housing to allow water to drip out. The motors and sensors can ordinarily stand some moisture with no ill effects; however, you will want to make sure the weep hole is aimed down. Install an accordian bellows-type rubber boot on the telescoping portion of the shaft, to keep water from leaking past the inner O-ring seal, and dripping down into the gears and sensor. Use silicon seal at all possible entry points. Don't do the final seal job until the dish is completely aimed, you will have to set the actuator mechanical-limit-switches first, and perhaps reposition the shaft on the mount, in order to reach the furthest Westerly and Easterly satellites. Undoing covers and unscrewing cable grommet nuts that have gooey sealant on them is no fun.

NOW AIM IT

Now that all the components are hooked up and the receiver is connected inside the house, it is time to aim the dish. If the dish elevation is set close to your site elevation, you have properly set the declination and focus, and you have the polar bar perfectly North and South (at night you will see it is pointed directly at the North Star), you will probably turn on the receiver, set it on scan mode, and find you already have it aimed at a satellite. If so, give thanks!

After your first exposure you'll be able to tell if you are getting downconverter hash and snow on the TV screen, and to recognize this type of snow is unique to satellite signals. If you have a blank screen, and no snow, you have a problem. See Chapter 17 for troubleshooting hints. But let's assume you have some semblence of a picture. Turn the hand-crank, or actuator, East or West to improve the picture. Adjust the rotor to improve it more. If you get a better picture, even just a watchable picture, that means everything is working. Move the dish to directly South, and find the satellite closest to direct South for your Longitude. (T2 in Chicago, W5 in California.) Adjust the elevation screw to improve the picture. It should need only a turn or two. (If your dish doesn't have an adjustment screw for elevation, throw it away and get one that does.) If you have a perfect picture, move the dish to a westerly satellite, such as G1 or F3. Get a fairly snowy picture and raise the dish, to see if it improves the picture. If it does, the polar mount needs to be turned, so the bottom lip of the dish is slightly more to the west. Loosen the set screws on the pin, and twist the dish slightly west, approximately a ½ inch or so. When you reposition the dish with the handcrank or actuator back to the east, to zero in on the same satellite again, the dish will have elevated itself a little, and you might find raising up or pushing down on the dish produces no further improvement. Usually, one or two corrections like this, should have the dish aiming optimized. All that is left, is instructing the customer on proper operation, warning them that it will take about three weeks to become accustomed to the types of programming, and operation of the hardware. Tighten things down, make a final inspection, seal everything up, and go home.

Another method of dish aiming is to set elevation on the most southerly satellite, then repositioning the dish to a most westerly one. Loosen the dish sleeve on the pin, swivel the dish east or west for optimum picture, then lock the sleeve down on the pin. A slight readjustment of elevation might be necessary again on the most southerly satellite, as a final tweek.

QUIZ

1. The trench containing the cables connecting the dish to the receiver must be at least _____ deep.

    a. ( ) 2 inches
    b. ( ) 6 inches
    c. ( ) 12 inches
    d. ( ) 3 feet

2. If protective conduit is not used for the entire cable run, it should be used in highly vulnerable areas, such as:

    a. ( ) between feedhorn and dish
    b. ( ) inside crawl spaces and attics
    c. ( ) under crushed rock drives and perimeter flower gardens

3. When splicing an underground TVRO cable, it must be realized that:

    a. ( ) the cable will be submerged in water at times
    b. ( ) trenches drain quickly and thus water rarely causes a short or corrosion problem
    c. ( ) the packed dirt will keep a taped splice secure and dry

4. If a 10 foot satellite dish is to be lifted, and set in place, three or four men could handle it by the edges, face up, and attempt to lower it onto the pin. If the pole extends to six feet above the ground, the dish has a depth of 2.5 feet, and the mount has a sleeve length of 12 inches, in addition to the six inch distance be-

tween the dish back, and the top of the sleeve, how high must the installers lift the outer rim?

a. ( ) 6 feet
b. ( ) 8.5 feet
c. ( ) 10 feet
d. ( ) about 3.5 feet

5. If the same dish is set on edge with the sleeve angled down, parallel to the face of the dish, and the men lift it vertically how high must the bottom lip of the dish be raised in order to slip the sleeve over the pin?

a. ( ) 6 feet
b. ( ) 8.5 feet
c. ( ) 10 feet
d. ( ) approximately 2 feet

6. A roof mount can safely be attached if eight long lag-bolts are securely embedded in the roof sheeting.

a. ( ) true
b. ( ) false

7. Weaker video reception, obtained slightly east or west of the dish boresight, and the prime signal of a satellite transponder is called:

a. ( ) boresight reflection
b. ( ) sidelobes
c. ( ) terrestrial interference
d. ( ) F/D ratio

8. It is generally a good practice initially to disregard the dish manufacturer's focus distance figure, and instead to adjust focus for best picture.

a. ( ) true
b. ( ) false

9. A dish eight feet in diameter and two feet deep should have a focal point of:

a. ( ) 16 inches
b. ( ) 24 inches
c. ( ) 32 inches
d. ( ) 64 inches

10. Most TVRO receivers, LNBs, LNAs, and coaxial switches, made and recommended for use with RG-6 coaxial wire, will not accept the #18 RG-6 center conductor.

a. ( ) true
b. ( ) false

Chapter 12

Terrestrial Interference

Terrestrial interference, or T.I., is a major concern for those in the TVRO business. It is enough of a problem that an estimated 10 percent of all dish systems are affected to some degree by it. For the installer-technician it is a threat. It can turn a profitable installation into a nightmare, for both the technician, and the dish owner. Learning how to avoid or overcome it is as important as learning how to aim dishes properly, or to install a roof mount correctly. If you don't understand it and are not capable of dealing with it when it is encountered you will constantly wonder if there are other, non-related problems. It is easy to waste time and energy, chasing the wrong trouble.

There is more than one kind of T.I. that can be encountered by TVRO dish systems. The major one emanates from terrestrial microwave communications relay towers (Fig. 12-1). These phone company signals carry voice communications, data, and video. Long before satellite communications came into being as we know it today, the FCC allotted spectrum space for fixed microwave communications. Later, satellites came into the picture and the same frequency spectrum was allotted to C-Band satellite transponders, in a sort of spectrum-sharing arrangement. In addition, microwave communications use a band in the 2 GHz range, as well as a 6 GHz segment. The main problem comes with the 4 GHz usage. The possible communications interfering carriers are frequencies listed in Table 12-1. As you can see, each C-Band satellite channel can have two communications transmissions operating only 10 MHz away from its carrier frequency. Since the satellite video channel actually uses a passband up to 18 MHz on either side of its carrier, the phone company carriers can, and are in its band. If the TVRO dish is in the path of a microwave tower relay signal, the TVRO system will pick up the phone company

Fig. 12-1. Microwave communications tower.

communications, which will appear on the video screen as pulsating noise. Figure 12-2 shows the bandpass curves of five satellite transponders, and the location of the communications carriers.

Since the TVRO dish is normally aimed at some southerly elevation, rather than facing directly across country at a relay tower, it would be a correct assumption that the phone company transmissions are being received at an unfavorable angle. The signal from the satellite is at a maximum, and the relay signal is less than optimum. The problem is that the TVRO signal comes from 22,300 miles away. The relay tower signal comes from perhaps 10 miles away! One hundredth, even one-thousandth of the relay tower maximum signal can destroy the C-Band signals. Generally, a relay tower signal less than −25 dB will be undetectable, and of no consequence. A signal above 0 dB will cause severe picture degradation, or complete wipeout.

Avoid It

The simplest solution to the T.I. problem is to avoid installing a dish in any areas in its path. This is a solution many TVRO dealers have chosen. To do this, you need to know the location of the towers in your area, and the geographic paths the transmissions take. If your firm covers many counties, this will not be easy. If you operate within a district with a 30 or 40 mile radius, you can find out pretty quickly where the trouble spots are.

Table 12-1. Possible Interfering T.I. Carrier Frequencies.

Satellite* transponder	Possible interfering carrier
1-(3720) MHz	3710, 3730 MHz
2-(3740)	3730, 3750
3-(3760)	3750, 3770
4-(3780)	3770, 3790
5-(3800)	3790, 3810
6-(3820)	3810, 3830
7-(3840)	3830, 3850
8-(3860)	3850, 3870
9-(3880)	3870, 3890
10-(3900)	3890, 3910
11-(3920)	3910, 3930
12-(3940)	3930, 3950
13-(3960)	3950, 3970
14-(3980)	3970, 3990
15-(4000)	3990, 4010
16-(4020)	4010, 4030
17-(4040)	4030, 4050
18-(4060)	4050, 4070
19-(4080)	4070, 4090
20-(4100)	4090, 4110
21-(4120)	4110, 4130
22-(4140)	4130, 4150
23-(4160)	4150, 4170
24-(4180)	4170, 4190

*Satcom/Comstar
24 transponder satellite

(courtesy of the Microwave Filter Company)

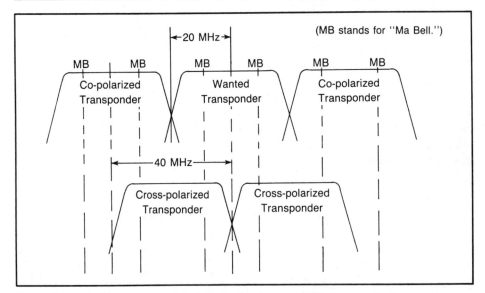

Fig. 12-2. Permissible microwave telephone carriers in the satellite transponder frequency.

SITE SURVEYS

Site surveys in unfamiliar locations should include a check for T.I.; if you take a dish setup to each site survey, you will quickly find out if T.I. is a problem. A much more efficient method is to use a Standard Gain Feedhorn, or simply connect up an LNA to a signal measurement device. Use it on site to attempt to locate T.I., and determine the origin directions by merely pointing it toward the horizon and moving it in a 360 degree pattern. For those dealers doing business in local areas, the T.I. trouble paths will quickly be recognized. Customers of other firms will tell you their experience with it. If their systems are not satisfactory, they will ask you to improve them. Once you recognize the trouble paths, you will be prepared to warn potential dish owners of the possibility of T.I. and the added costs it will entail.

The big problem to you isn't the T.I., it is the surprise that T.I. gives you and your customer. If you've made a deal and given your best price to get the sale, then find the site is going to require upwards to $1,000 or more in filters to obtain a satisfactory signal, either you, or your customer is going to have to pay. It is too late after the pole is in, the concrete set, holes drilled in the carpet and house, and the equipment installed, to cancel the contract. It's too late to ask the customer to pay another $1,000, unless he is the type that likes surprises!

To partially solve the problem, you should print a T.I. exception clause on your sales and advertising forms, so that in the event of a T.I. encounter, you are legally covered, and not bound to give away expensive filters. It's best though to contractually alleviate this problem in advance.

Other Interference

While relay-tower C-Band T.I. is the main problem, you might find 2 GHz relays troublesome also. Broadcast station video links splattering all over the countryside, can be your enemy. In rare instances, other out-of-band frequencies, such as fixed- and mobile-amateur radio/TV (1710 to 2290 MHz), and other bands can conceivably interfere. A reason for having perfect F-connections on your shield wires is to eliminate pickup of T.I. by the wiring after the signal has passed through the downconverter of your system.

Identification

The easiest way to identify T.I. is to look at the picture, and recognize the transmission visually. If it is T.I., and shows up as an ABC multiburst test pattern, you can find out the source and frequency, and attempt to trap it out. If it is communications relay hash blotting out certain channels, you can sometimes tell what the frequency is, and order filters to eradicate it. If it is affecting only one polarity of channels, and only a few of them, you can tell what the frequency is.

The most certain way to identify it, is with a spectrum analyzer. These oscilloscopes display a broad band of frequencies, and allow you to select specific bands to inspect visually, pinpoint the T.I. to its exact frequency slot, and allow you to decide on a course of action to take. Microwave Filter Corp., 6743 Kinne St., East Syracuse, NY 13057, makes filters for each application. They also publish a most helpful catalog you can order. If you are unable to solve your T.I. problem, MFC engineers will assist you, even contracting to visit your site to analyze and correct the problem. If you are a member of your local or state satellite technicians association, you will probably have one or more members who are experts in T.I., and will contract to help you.

In the most frequently encountered cases the relay tower will be sending you six separate 4 GHz carriers from the nearest tower. They will be 80 MHz apart. Fortunately, they will all be of the same polarity. When you find a problem on every 4th channel it is an indication of a phone company relay tower problem. Unfortunately, if the T.I. signal power is too great, only a few interfering carriers can overload the receiver electronics so that most channels are affected. So what you do depends on the nature and size of the T.I. problem.

If you have a mild case, T.I. signals −6 to −12 dB or lower, you might find it insignificant enough, that watchable pictures are available on all satellites, and only a few channels, on one or two satellites, have any noticeable intermittent hash in the picture, or fuzzy pulses in the sound. An explanation to the customer may suffice, and be the best answer. Larger amounts of interference require action on your part.

Notch Filters

Notch filters are merely hi-Q LC traps, that insert into the 70 MHz *if* line at the receiver. They are preset to 60 or 80 MHz, which is 10 MHz off from your 70 MHz channel center frequency, and directly on frequency for communications type T.I. This might be your first attempt at combating the problem. If it works, great. You have done the job for a cost in the $100 range.

A 60-80 MHz notch filter that is tunable, is also available. It allows you to zero in exactly on the offending carriers. Others are available that switch either the 60, or 80 MHz trap in, if and when either is needed. Since these filters eradicate a portion of the 36 MHz passband used in each channel, and therefore some of the picture information, it is best to filter only what is needed, and be able to switch out the filter on other channels.

Bandpass Filters

If the notch filters don't solve the problem, your next step is to try a bandpass filter on the 70 MHz line. Instead of suppressing frequencies in a narrow band at 60 and 80 MHz, it chops off either end of the passband. This could deteriorate the picture even more, but can be what is needed to save the installation. These bandpass filters are available from MFC as noted previously, and from ESP of Lexington, KY, plus others. Some receivers have been produced with built-in filters that can easily be inserted or removed, as needed.

AFC

Automatic Frequency Control circuits are designed to lock in on strong signals, and to hold the receiver on that frequency with no drifting. Since the T.I. signal is near the channel frequency, there is no reason the receiver should not lock in on the T.I., if it is stronger, rather than the desired channel carrier. Some receivers have an AFC defeat-switch, in the off position it can allow you to keep the T.I. signal from causing the AFC to shift the received frequency down, or up to the T.I. frequency. Quartz-locked tuning receivers, without AFC, can simply lock in on the T.I., and compound your problems. Substituting a non-AFC receiver might help you.

Block Systems

In block systems with no 70 MHz loop, all 24 of the channel signals from the dish are traveling into the receiver at the same time. Rather than trapping one side or the other of the 70 MHz *if* carrier, that takes care of all the channels the receiver can tune in, block systems might require individual traps for each offending carrier. These are available, and you can create a waveguide with large numbers of individual resonant cavities, each tuned to short out offending frequencies. The problem with using multiple waveguide traps is the cost. If $2,000 to $10,000 is required to do the job effectively, it will be bail-out time for all but commercial, or very special cases.

While having your system hooked up to a spectrum analyzer, a T.I. diagnosis kit can be used to pinpoint the frequencies bothering the system. This cavity type waveguide allows you to hook your LNA output into the kit waveguide, and then connect the signal on to the downconverter. The MFC 4043A is calibrated so that you can see exactly what frequency trap produces a decrease in T.I., or eliminates it. After you have located each unwanted carrier, fixed traps are ordered.

Shields

If all else fails, some interference can be reduced by erecting screen shields between the T.I. source, and the dish face. These extreme methods might not be acceptable to some dish owners. Some improvement can be made by selecting a dish location shielded by a structure that has metal-backed insulation in the walls, aluminum siding, corrugated walls, or masonry. Selection of a ravine, or ground depression, for the dish location might help. We have one 10 foot dish in our town that was moved from the front yard location, and placed practically up against the east side of an aluminum-siding house. It solved the problem with T.I. It is common to see T.I. reduction take the form of shields built perpendicular to the dish face, around its rim. These can cause some change in the lobe response of the dish, but if they eliminate or reduce T.I., obviously they are worth it.

Cancellation

A method used in many other cases of interference, or unwanted signals is called signal cancellation. In the TVRO instance, take a (SGF) Standard Gain Feedhorn and LNA, or another dish setup, and aim it at the offending T.I. source. By inserting the output of this LNA back into the output of the system's downconverter, the T.I. will be opposite in signal polarity, or phase, and will cancel. By adjusting the signal level of the SGF-LNA to exactly the level from the downconverter, the T.I. will be reduced or eliminated, and the satellite signal will be virtually unaffected. Figure 12-3 illustrates phase cancellation.

Ku Band

Since communications frequencies are not at the Ku band TVRO frequencies (11.7 GHz to 12.2 GHz), you can install a Ku band system and solve the problem. Some modern receivers are already set up for Ku band, and all that is needed is to change the LNB, and replace or refit the existing feedhorn with a Ku band horn.

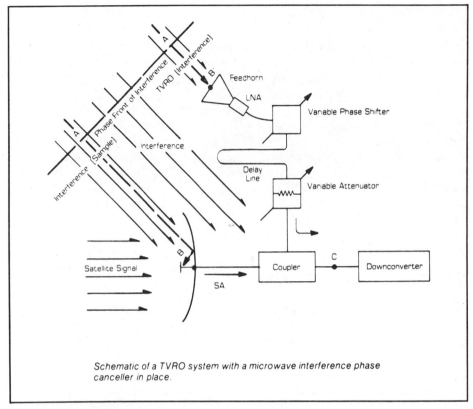

Schematic of a TVRO system with a microwave interference phase canceller in place.

Fig. 12-3. Drawing of microwave interference phase canceller.

QUIZ

1. A microwave communications antenna tower is always receiving signals from another tower as well as transmitting to one or more towers.

   a. ( ) true
   b. ( ) false

2. AT&T, and other phone company towers will always be transmitting voice communications only.

   a. ( ) true
   b. ( ) false

3. The C-Band T.I., caused by microwave communications, carrier frequencies that are about 20 MHz either side (higher and lower) of the satellite channel carrier frequency.

   a. ( ) true
   b. ( ) false

4. Which of the received T.I. signal levels would cause the most deterioration to a satellite signal?

   a. ( ) 10,000 microvolts
   b. ( ) 2 dB
   c. ( ) −20 dB
   d. ( ) −40 dB

5. If the T.I. arriving at the dish site is strong, it can be detected by merely aiming an operating LNA or LNB in its direction, using no reflector.

   a. ( ) true
   b. ( ) false

6. A spectrum analyzer has a display oscilloscope screen, and can show a frequency spectrum with a maximum bandwidth of 500 kHz.

   a. ( ) true
   b. ( ) false

7. T.I. emanating from a TV station video link tower, broadcasting in the 2 GHz band, might appear on the satellite receiver screen as:

   a. ( ) random noise
   b. ( ) herringbone pattern
   c. ( ) pulsating video pattern

8. A filter used to reduce the effects of T.I., that eliminates both extremes of the channel passband, leaving only about 16 to 18 MHz of band centered on the satellite carrier, instead of the desired 36 MHz wide passband, is called:

   a. ( ) a notch filter
   b. ( ) a bandpass filter
   c. ( ) a standard gain filter

9. A T.I. filter that uses negative feedback, and operating electronic circuitry, is called:

   a. ( ) an active filter
   b. ( ) a passive filter
   c. ( ) a waveguide filter

10. Ku band satellite channels do not have other microwave communications sharing the band as C-Band does.

   a. ( ) true
   b. ( ) false

Chapter 13
Selecting the Best U/V Antenna

What antenna should you install for uhf/vhf TV reception? It might seem that question should be a simple one. Years ago in Indianapolis, my firm installed what was called an Indy Special antenna. It was all we ever used. The first Indy Specials were composed of two bays. One bay or section was designed to be headed toward the north side of the county, and to receive channels 6 and 13. The other section was cut for channel 8, located southeast, and channel 4, about 20 miles south. Figure 13-1 shows a typical area special antenna.

This antenna was cheap. It could be mounted with the included tripod, the entire installed job sold for $40 to $60. In many areas of the city it worked fine. If you were located between the channel 4 and channel 8 towers, chances are that ghosts were a problem, since you couldn't point the ⅛ section in two different directions. Later the Indy Special was designed as a 3-section antenna. One section was cut for channel 4, 8, located southeast, and channel 4, about 20 miles south. Figure 13-1 shows a typical area special antenna.

This antenna combination isn't the best antenna for the Indianapolis area, but considering it was lightweight, could be mounted on a roof by one man in a matter of minutes, and the price was so affordable, that's what was sold by area antenna and television dealers.

These multiple-section specials have been designed and manufactured for most metropolitan areas. They work much better than rabbit ears. If the location is just right, they are absolutely the best choice for an antenna. By including a director on the front of each section, and a reflector to the rear of the active element, the section is direction-

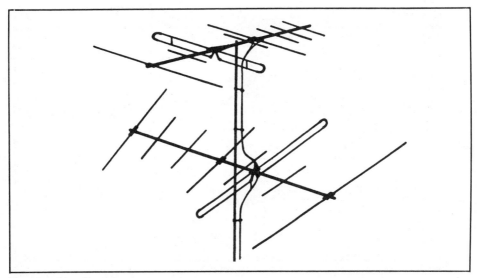

Fig. 13-1. Area Special antenna.

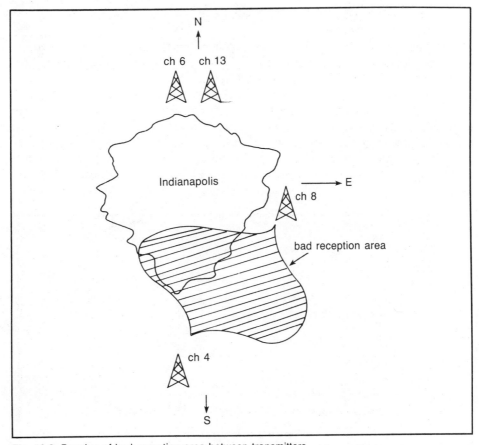

Fig. 13-2. Drawing of bad reception area between transmitters.

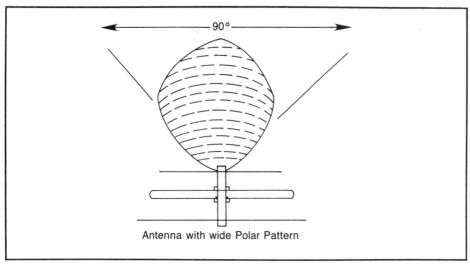

Fig. 13-3. Wide polar pattern antenna.

sensitive to some degree. Figure 13-3 illustrates the wide polar pattern of a multiple-section antenna.

Directionality is more desirable in a metropolitan area than gain. The antenna needs to reject signals from the sides and rear in order to reduce the reflections that are often a problem in urban areas. The antenna sections need to have a narrow frequency response, if possible, so that only the desired channels are received on each section.

To achieve a narrow frequency bandwidth, the elements of the sections have to be extremely thin. The thickness of a clotheshanger would be desirable. Figure 13-4 plots a desirable frequency response for an antenna.

It would be difficult to manufacture a wind-resistant antenna with very thin elements economically. Such thin elements would cause the characteristic impedance of the antenna to increase above 300 ohms, so it isn't practical to use that design. The common design

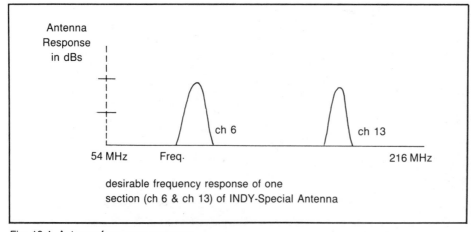

Fig. 13-4. Antenna frequency response.

then, is a result of a host of compromises. As a community becomes familiar with the cost and results of an antenna, the perception of set-owners is such that, an antenna installation assumes a certain perceived dollar value. In the 1950s and 60s the value was around $50 to $60 in Indianapolis. In some locations this combination is worse than rabbit ears. In other locations, one or two of the four available vhf channels might be received well, and the others suffer from severe ghosts. While the area special antenna combination can serve 40 to 50 percent of the community reasonably well, and carries a bargain price, it should not be relied on as the only rooftop antenna to be sold or recommended.

THE YAGI—A BETTER SOLUTION

While the area special antenna sections are described as modified Yagi's, they are more of a modified selective band antenna. Yagi's are generally designed to receive only one channel. They are called cut-channel Yagi's. (See Fig. 13-5.)

In instances where the single antenna cannot be aimed in a direction to receive all the available channels, several Yagi's can be used. Note in Fig. 13-5 that the Yagi has several more directors than the area special sections. The directors and reflectors are also very nearly the same length. The directors can be about 95 percent as long as the driven element, and the reflector can be up to 10 percent longer than the pickup element.

The many directors and reflectors do two things: They increase gain, and reject signals coming from any direction except the front end. Figure 13-6 plots the polar pattern of a Yagi antenna.

In metropolitan areas you will find antenna arrays consisting of Yagi's for each desired channel, mounted on schools, apartments, hotels, cable towers, and anywhere else that a clean signal with no ghosts is needed. The idea is the same as with the area specials— to aim a single antenna at the broadcast tower, and to reject reflections coming from the sides and rear. The array also should reject signals of a different frequency, coming in from the front.

Because of the more numerous elements, and generally heavier construction, Yagi's are more expensive than area specials. Instead of an antenna array having two or three

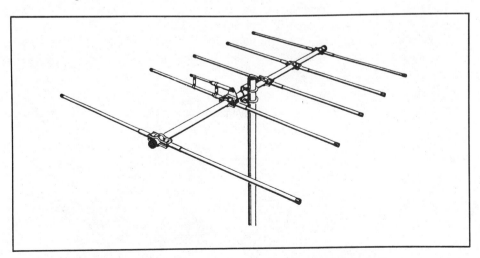

Fig. 13-5. Yagi antenna.

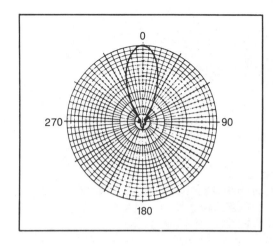

Fig. 13-6. Antenna polar pattern.

sections like the Indy Special described, four Yagi's would be used in Indianapolis. Now that three stations have been added, the Indy Special or the Yagi array would need an additional uhf broadband section. Fortunately, the three uhf broadcast towers are close together geographically, so area residents have little problem receiving uhf.

Where the area special reception is plagued by ghosts from nearby water towers, tall buildings, power lines, and other reflective surfaces, the Yagi array can be the best solution. Since the rotatable antenna is out of the question for multiple family dwellings, the Yagi array is the only answer for them.

To further improve the signal rejection capabilities of the Yagi array, each antenna section can be channeled through a bandpass filter that is responsive only to its frequency. This will eliminate some of the adjacent channels from interfering with those desired. If a strong channel still overlaps the other channel reception despite the use of Yagi's and filters, the next step is to use phase cancellation to cancel the unwanted signal electronically.

Phase cancellation of ghosts and unwanted channels, the use of screens to block out reflections, and other drastic steps taken to produce a clear signal in a problem location, are remedies that are rarely used. This is due to the high costs of extra materials, and experimentation. Knowledge of how to solve these difficult problems, for those cases where price is no object, is desirable for any antenna installer-technician.

ROTORS

You have probably concluded that the simple solution to many urban area problems, where the broadcast antennas are in different locations, is to install a rotor with a highly directional all-band antenna. You are correct. Such an installation will not completely eliminate reflections, but will generally minimize them to where the picture is tolerable.

The problem with rotors in metropolitan areas is cost. You are competing with built-in rod antennas, rabbit ears, and cheap area special installations. The rotor is going to require $100 in retail costs including the wire, plus the mount must be stronger, and requires some additional pole length. In addition, the low-priced area special elements should be replaced by a single narrow beam width antenna. This can add another $50 plus extra installation time. The rotor installation can be a $300 job, compared with $2.95 for rabbit ears! The professional technician/salesman will constantly be attempting to sell customers on the very best antenna system for the location. When offered quality, many set owners will opt for a rotor and a proper MATV distribution system.

THE FRINGE

As the technician works further from the metropolitan area, the problems and solutions are different. Twenty to 30 miles away from the broadcast stations the problem might be one of distance. Simply using enough antenna to produce a 0 dB signal is the prime objective. Further out you will probably find it necessary to use preamplifiers. Even further you might be trying to get signals from more than one city.

Where only one direction is required, a stationary antenna, with a broad polar pattern, could be the best answer. In the Indianapolis example, with the stations being 25-30 miles apart, a resident located 10 to 20 miles either east or west of the city would need a polar pattern that allows signals 45 degrees off-center to be received nearly as well as those straight ahead. The antenna would be pointed in a compromise direction, between the towers. Figure 13-7 is a drawing of a location 20 miles west of town.

An obvious improvement would be to add a rotor. An amplifier is not needed because the signal strength is sufficient at this distance, unless the location is behind a hill, or multiple receiver hookup is required. The rotor will allow zeroing in on any of the three locations. By installing a rotor and a narrow beam antenna, you might find the customer won't take the time, or doesn't know how, to aim the rotor. So they leave it in a position that is watchable, but far from optimum. A wideband stationary antenna would perhaps be better for this type of family. As the quality expectations of people grow, more and more will want a rotor for this situation.

MULTIPLE CITIES

In Fig. 13-8 the most common antenna arrangement is a rooftop antenna, pointed toward the prime reception city. In this case Indianapolis is about 50 miles away. Without a tower or an amplifier, the fringe area signal is rarely 0 dB or above. The majority of homes were sold double-bay conical antennas years ago. Double-bay conicals produce exceptional vhf signals, when their low cost is considered. These eyesore arrays work

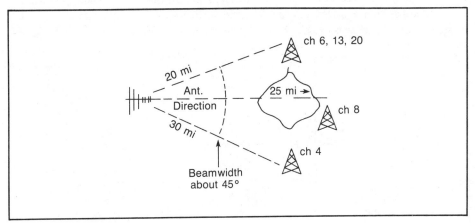

Fig. 13-7. Antenna reception pattern.

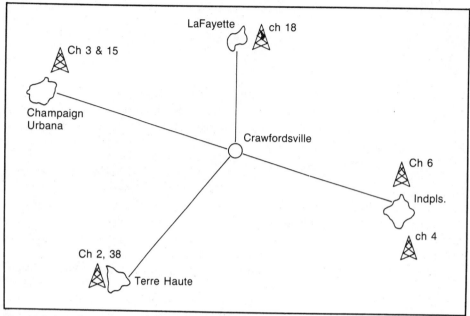

Fig. 13-8. Location between 4 broadcast areas.

for channel 2 through channel 13 because of their construction, which resembles the letter X, or a bowtie as the front, or driven elements. That mass of non-resonant metal waving around above the chimney produces a vhf signal of higher strength than most all-band antennas. The X-configuration of the driven elements produces wide bandwidth. Double-bay conicals were not satisfactory for uhf stations, and few homeowners could receive them. Figure 13-9 illustrates a typical double-bay conical antenna.

This ugly conical antenna works great for near-fringe, and local areas; it also works well in areas where strong rear-direction signals are available. The reflectors do not reject rear-direction signals well. In those unique locations 30 to 50 miles between two vhf locations, they are the best low-priced antenna. It is rare that you can add a uhf section to a double-bay conical array, and have as good performance from the uhf antenna. That is because the uhf signal is usually weaker due to its line-of-sight characteristics of signals,

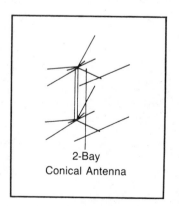

Fig. 13-9. Drawing of 2-bay conical antenna.

foliage, blocking, and atmospheric moisture diffusion. This presents somewhat of a dilemma for technicians. Replacing a broken down conical with a technically superior all-band U/V antenna can frequently leave you red-faced, as your $150 antenna produces less signal on favorite vhf channels than the one the customer originally had.

In Fig. 13-8, if the customer wants to improve reception, a rotor is most desirable; not only are there vhf stations at opposite directions, a nearby local uhf, but also the signals from Terre Haute. While it is possible with a stationary antenna to get watchable pictures on the two uhf and vhf stations in Champaign, plus the seven from Indianapolis, little, if any, signal will be seen from Lafayette or Terre Haute. Nine stations could be sufficient for the soap-opera and game-show crowd, but technicians should attempt to introduce viewers to all available channels. In this location 13 channels are available with the use of a rotor, wideband antenna, and preamplifier.

UHF ONLY

Some cities have uhf stations only. The most practical uhf antenna is the 4-bay type with screen reflector, as shown in Fig. 13-10. Measurement will show that it provides more gain than most U/V combination antennas. It does this because it is, in effect, four stacked uhf receptors with a reflector. Just as the conical is wide banded due to the X-shape of the driven elements, the similarly designed elements on the 4-bay, about 12 inches long, will capture channel 15 just as well as channels 5, 30 and 60. The reflector adds a measurable boost. It also eliminates rearward reflected signals. Because uhf generally requires more gain at the antenna than vhf signals, the 4-bay uhf is a natural. The Winegard KU-420 is one version, Channel Master and General Instruments make similar models. These produce nearly double the uhf signal strength of combination U/V antennas.

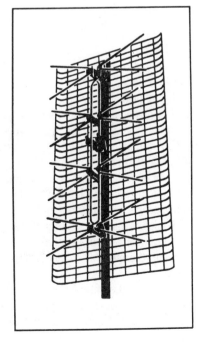

Fig. 13-10. Uhf 4-bay antenna.

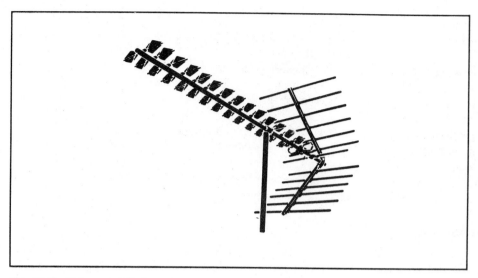

Fig. 13-11. Uhf corner-reflector antenna.

In UHF-only areas, the KU-420 might not be adequate if ghosts are a problem. In that case a uhf antenna designed with multiple directors is better. The more directors, the narrower the beamwidth, the better antenna rejects off-boresight reflections. Figure 13-11 shows such an alternate construction.

In some few locations where uhf signals are available from opposite directions, the screen reflector on the KU-420 type antenna can be removed, allowing rearward signals to be received too. Frequently, the reduction in gain, and the increased susceptibility to ghosts negates this approach.

Fringe

The all-band U/V long-range antenna is the most used unit in rural areas. Gain on all three bands of 12 to 16 dB is possible. The larger arrays have an element cut for each vhf channel. This provides an even response over the entire vhf Hi and Lo bands. The front-end uhf section has numerous directors, and either a bow-tie or folded dipole active uhf element. This gives broad-band capability to the uhf frequencies. Be careful of U/V antennas that use only one rod-type uhf active element as the high end, channels 40-83 could drop off too quickly.

My experience is that a combination of the KU-420 type uhf section, and a separate vhf-only section, fed into an amplifier with 17 dB for vhf, and 24 dB for uhf, is superior to any single combination antenna setup. A 6 dB improvement in uhf antenna gain, plus 7 dB higher gain in the amplifier is hard to match using any other antenna. If you are working with a tower or other location where the uhf is not lower in power in comparison with vhf, the single all-channel antenna is sufficient, easier to install, and provides adequate signal.

Some of the antenna manufacturers provide specification sheets for their antennas. The polar pattern and grain-graphs are useful to the technician in selecting the best antenna for a particular location. In addition to these, the choice can also be influenced by the

Table 13-1. Antenna Model Numbers.

Brand	Model	Usage	Elements	Gain V,Vh,U	Retail $
General instruments	VU-937	Deep fringe	64	8, 13, 13	200.00
Blonder tongue	YH-210	CH 2 YAGI	10	13.6	N.A.
Channel master	3667	Suburban +	27	N.A.	$50
Winegard	KU 420	UHF	4	N.A.	$40
Winegard	TV 5030	VHF fringe	17	N.A.	$62

physical design. Some antennas are easier to unfold and put together. Some have more sturdy mast-mounting methods and will resist wind better. Newcomers in the antenna business can get help in selecting the right antenna from their electronic parts distributor. He often has input from other area installers that have experience with the antenna brands available. The physical appearance of your installation could become a factor in customer recommendations, and your future business. Selection of all the components of an antenna installation with an eye toward an aesthetically pleasing appearance can be valuable. Sometimes the old installation, poles, or tower, is unsightly. You might be able to convince the customer to change the method of mounting, to get rid of the bent or rusted poles or tower, and to install the new antenna on a less visible part of the house.

Some antenna model numbers with suggested usage are listed in Table 13-1.

QUIZ

1. Many homes are still using local area antennas originally installed to receive vhf only.

   a. ( ) true
   b. ( ) false

2. Simple area special antennas, made for urban areas, are a problem because . . .

   a. ( ) gain is too high
   b. ( ) a rotor is needed with them
   c. ( ) coax cable cannot be used
   d. ( ) directability is not right for many areas

3. To reduce ghost pickup or reflected signals, use an antenna with a . . .

   a. ( ) narrow polar pattern
   b. ( ) wide polar pattern
   c. ( ) figure 8 polar pattern
   d. ( ) narrow reflector screen

4. In a location which has the desired TV transmitters located 45 degrees either side of the boresight direction of your antenna, that antenna should have _____ directors.

   a. (   ) many
   b. (   ) few

5. A cut-channel Yagi is a _____ band antenna.

   a. (   ) narrow
   b. (   ) wide

6. Even though you are using a Yagi antenna for channel 4, a strong channel 5 station causes some interference. What can you do to reduce the channel 5 interference?

   a. (   ) change to twin-lead transmission wire
   b. (   ) install a bandpass filter which rejects all but channel 4
   c. (   ) install a channel 4 trap
   d. (   ) use a conical antenna instead of a Yagi

7. Even though a rotor and a proper antenna is the best way to minimize ghosts and increase signal, many metropolitan area people do not invest in them because . . .

   a. (   ) cost is too high
   b. (   ) city ordinances do not permit them
   c. (   ) they are unsightly
   d. (   ) guy wires must be used with them

8. A conical antenna with pickup elements cut to the length of channel 6 will also pick up channel 2 and channel 13 if . . .

   a. (   ) the elements are thin and in a horizontal plain
   b. (   ) the elements are thick and shaped like a bow tie
   c. (   ) front and back sections are electrically tied together

9. While the 4-bay screen reflector uhf antenna has higher gain than combination antennas, it is not as directional because . . .

   a. (   ) the combos have several directors
   b. (   ) the corner reflectors used on combos are more efficient than the screens on the 4-bay
   c. (   ) the 4-bay active elements are in the shape of a bow tie
   d. (   ) 4-bay antennas cannot be used with rotors

10. If signal strength is too low with a certain antenna having a gain of 12 dB, the best way to increase signal strength by 12 or more dB is to:

a. ( ) use a 24 dB antenna
b. ( ) add a signal amplifier
c. ( ) change to coax wire
d. ( ) raise the antenna 10 feet

Chapter 14

Antenna Rotors

Few antenna installations would not benefit from the use of an antenna rotor or positioner, yet less than 10 percent of homes using rooftop antennas or towers have them. Rotors add a hundred dollars or more to the cost of an installation, the prime reasons more aren't used. People who have never used one really don't know what they are, or how they work. Some feel they are too complicated. Still others have a rotor, but rarely reposition it, choosing instead to watch whatever station at which the antenna is already aimed, or settling for a less-than-optimum picture, simply because they don't want to bother with turning another gadget. It isn't an easy sales job to explain verbally the superior picture quality often possible when the antenna is aimed directly at the station. Some people think a rotor might just be something else to go wrong, and after all, who wants to have to repair something so high above the roof—especially in cold weather? With the advent of cable TV and home satellites it would seem the demand for rotors would be diminishing.

Surprisingly, rotor sales are alive and kicking. The $100 price for a rotor doesn't seem to be such a large amount, as it did 10 or 20 years ago. It is only two or three months cost for cable TV. It's about one-fourth of the price of a TVRO descrambler. It is only a little more than the price of an amplifier or antenna. For its utility and dependability the antenna rotor can be one of the best bargains in TV.

NEW REASONS FOR HAVING A ROTOR

Another factor adding to the popularity of rotors is that in the past 15 to 20 years a lot of new TV stations have begun broadcasting. Dozens of new uhf stations, often

Fig. 14-1. Rotor mounted on 10 foot pole.

located in smaller cities, have become available. A majority of the population has signals available from more than one direction. Frequently it is desirable to turn the antenna to three or more different directions. Someone living in Greensburg, Indiana can easily receive Indianapolis, Richmond, Cincinnati, Lexington, and Louisville channels if they have a rotor and amplifier.

Customer expectations have changed the public attitude about TV programming. It is quite evident, when you install a TVRO system to find that the customer also wants his broadcast reception antenna improved. After 10 years of being content with one or two snowy, ghosty channels, you might think the broadcast stations would be of little interest, or that any additional expense would be justifiably deferred especially with over 100 satellite channel choices. Quite often, as TVRO dealers will attest, the customer today seems to have a new perception of the value of TV, and decides to go ahead and upgrade the rooftop antenna system. This frequently includes finally hooking up the other rooms in his home with a mini-MATV system and rotor.

Rotors are not difficult to install. The Channel Master 9510 requires a 3-wire cable for the ac power, for the small phonograph-type ac motor that drives the plastic gears and turns the rotator in a 360-degree pattern. On most jobs a 10 foot galvanized pole can be cut in two, then one half used as the base or mount pole and the other as the movable pipe holding the antenna. Figure 14-1 shows such an installation.

POTENTIAL PROBLEMS WITH ROTORS

- On towers and difficult-to-reach poles, the extra weight requires stronger antenna technicians.
- Adding the rotor wire can require drilling more holes in the entry path if the existing system left no extra feedthrough space.
- Frequently the orientation instructions don't match the rotor and one has to climb back up, loosen the U-bolts and reorient the antenna to match the control-box indicator.
- Dead-on-arrival rotors are common. Control boxes don't work and the dials slip. Motors jam or just quit.
- The rotor adds several pounds more weight, and generally requires some added pole length, which can mean a stronger mount is needed. Where a stationary

antenna could be eave-mounted, the rotor and extra pipe can cause too much wind leverage.

- When servicing older rotors, the wiring sequence for four- or five-wire units might be unknown, with no legend to guide you, and no numbering system to assure the wires are in the correct sequence.
- Older rotors are constantly in the weather. Attempting to replace a pole or move the rotor often results in bolts broken off flush with the rotor case.

While the list might seem to point out that rotors are troublesome, and perhaps an item to stay away from, actually they are not difficult. Most work well for 20 years or longer. They are not as prone to lightning damage as amplifiers and TV sets. Older unpopular models that have served for 10 or more years are candidates for replacement, if the poles or antenna is to be changed. The results of installing a new rotor, for the first time, in terms of picture quality, often make an antenna technician appear as some sort of hero to the family. The competition in this business is not great; rotors, wire, and installation can be profitable.

Automatic rotors that stop once the antenna reaches the dial setting location, are so inexpensive that it isn't practical to install manually operated types. A rotor will usually travel a full 360 degrees in about a minute. Most can be depended on to handle an antenna pipe up to 10 feet in length.

Where a heavy antenna is used, the rotor should be assisted in handling the extra weight and wind leverage. Thrust and alignment bearings are available to allow you to concentrate the mast weight on the bearing, either above or below the rotor; this allows the rotor to change direction without any weight being directed on it, and very little side leverage. For those who like to install guy wires, most rotors are installed above the guy wire connecting ring, so the pole has freedom to move. Guy wire rings with a built-in bearing are available too, allowing higher guys above the rotor, and a more secure installation.

Towers are usually manufactured with the expectation of rotor use. Some have a flat plate in the top section on which the rotor can be placed, slightly below the guide pipe at the peak of the tower that is used for inserting the antenna pole. Rohn and other tower manufacturers also supply brackets that can be attached nearly anywhere on a tower leg, and can be swiveled to position the rotor, so that its rotating antenna pipe does not bind in the guide pipe. The guide pipe has set-screws that also allow a short pipe atop the rotor to hold the antenna if no rotor is used. The rotor can be set on top of the tower on this pole, then another short pole mounted on top of the rotor to hold the antenna. The most secure method is to mount the rotor a foot or more below the tower guide pipe, and to install an alignment/thrust bearing in the guide pipe, to handle the sway and the pipe/antenna weight. Figure 14-2 shows a rotor being installed.

Repair of rotor hardware is not the most rewarding service job you will find. I won't explain the control box and actuator motor wiring and gearing here. The most practical repair is replacement of the entire unit. This approach is the most common, even though the actuator usually is at fault. After a time working with rotors, you will acquire a number of the old replaced units, that will have one good component—the controller or the actuator. There will later be instances where you will want to "save-the-day" by solving a customer problem, using one of the used units. Of course, using a *known* good used

Fig. 14-2. Rotor located in top section of 50 foot tower.

unit for troubleshooting is helpful too. Having a used, reconditioned rotor available on your sales floor for do-it-yourselfers, who could be headed for the discount store is a good alternative.

As with any other SAM hardware item, becoming familiar with a product like a rotor, learning the best ways to install and service it, to aim it, and to recognize problems, is a key to being able to tackle each new installation challenge that comes up efficiently, and profitably.

QUIZ

1. Most rotors will turn about 2 and ½ revolutions or circles before hitting an automatic stop.

   a. (   ) true
   b. (   ) false

2. Usually, an inoperative rotor will be found to have both a bad actuator, and a bad control box.

   a. ( ) true
   b. ( ) false

3. Which type of bearing is meant to absorb the weight of the antenna pipe in a rotor installation?

   a. ( ) thrust bearing
   b. ( ) alignment bearing

4. Which type of bearing is meant to absorb side to side sway of the antenna and pipe?

   a. ( ) thrust bearing
   b. ( ) alignment bearing

5. An alignment bearing will usually be mounted . . .

   a. ( ) around the pole between the antenna and rotor.
   b. ( ) below the rotor in most cases.

6. An antenna, installed 15 years ago in a system using a rotor, is bad. A likely problem in replacing it will be:

   a. ( ) the U-bolts holding the antenna will be
   b. ( ) the rotor studs will twist off.
   c. ( ) the pole will be rusted and not strong enough to reuse.
   d. ( ) all of the above

7. Rotors use dc power to energize the actuator drive motor.

   a. ( ) true
   b. ( ) false

8. Guy wires must always be mounted to the stationery position of a rotor, or below, so that the antenna pipe can be rotated.

   a. ( ) true
   b. ( ) false

9. A rotor can be mounted on a short pole, attached to the tower guide pipe.

   a. ( ) true
   b. ( ) false

10. Considering the drastic difference in cost between a manually-operated and an automatic rotor, most people purchase the manual type.

    a. ( ) true
    b. ( ) false

Chapter 15

Antenna Towers and Mounts

Antenna towers are intimidating to most people. It's one thing to install a chimney mount on a flat roof house, and quite another to climb a 40 foot tower. The faint of heart quickly decide tower work is not their cup of tea. It is because of this fear of heights and unhandy working conditions, that tower firms have little competition, and are profitable. If you won't go up on a tower and replace a broken antenna, then you had better be prepared to pay someone else a lot, to do it for you. Figure 15-1 shows a tower installed without a cement base.

Tower work takes strength. Working with a lineman's belt, and carrying all the tools you need to install an antenna, rotor, amplifier, and coax is unhandy. Each move seems to require about twice the effort that it would on the ground. The tower itself is always in your way, the antenna pole is not strong enough to support you, so you can't get up quite high enough. The tower rungs hurt your feet, and the new antenna can't be slid up the mast and turned into position to clamp it without bending at least two elements. Dropping a U-bolt or your electrical tape can be a cause for major concern.

Fortunately, you don't have to do any tower work *too* difficult for you or your men. There are other companies who specialize in towers, who are glad to work with you to do all of the tough jobs. All you have to do is to price the job properly, call the tower company, and stand by to hand them the components, while you stay on the ground.

If you do get heavily into the SAM business you will find that you can handle a majority of the tower problems. Training a young weight-lifter to do the heavy work could be the way to go, if you are not capable of this type of effort.

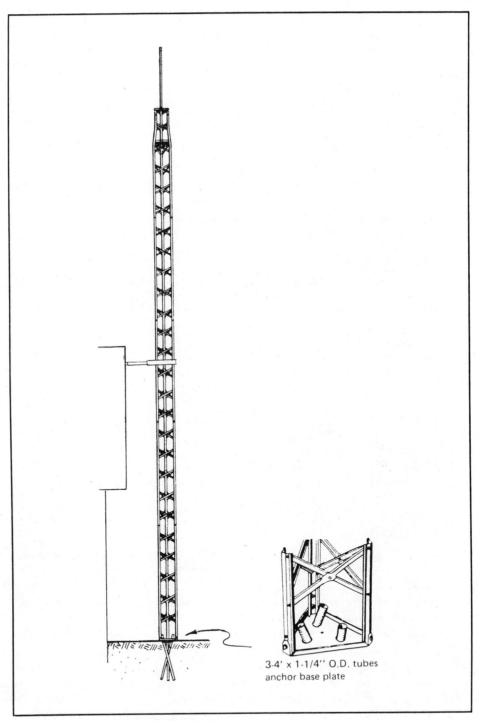

3-4' x 1-1/4'' O.D. tubes
anchor base plate

Fig. 15-1. Tower bracketed to house gable.

The third way to get around the hazardous-heights problem is to simply turn down tower work. There is plenty of roof-top and MATV work. Most SAM businesses will not choose this option, since the towers are a part of the business, and are not as difficult as they seem.

NECESSARY EQUIPMENT

With the wind blowing, and a tower swaying, it isn't hard to visualize it collapsing under your additional weight. The simple solution to this is to climb up 20 feet or so, (higher on an extremely high tower) and attach some guy ropes. Tie them off while you are working, and the tower can't bend over. Obviously, a tower that has completely rusted out, or an aluminum tower that has long since lost its strength, should be taken down and replaced.

To keep from falling, you must use a lineman's belt. It costs about $200; however, there is no way to use both hands on a tower without one. It keeps you from falling backward and allows you to rest against the wide, flat, strong, leather belt. You also need construction work boots, not tennis shoes, and an electrician's belt to hold tools. You will need the following tools for most jobs:

- Dykes
- Pliers
- $7/16$-inch box wrench
- $7/16$, $1/2$ and $9/16$-inch ratchet sockets and drive
- PVC tape
- Screwdrivers, both blade-type and Phillips-head
- Crimpers and "F" connectors
- Vise-grips
- Box of assorted nuts, washers, bolts, and U-bolts

It is difficult to haul a 60 element antenna up a tower. The answer is to crawl up yourself, then hoist the antenna up slowly by its end with a rope. It is helpful if an assistant is guiding the antenna with a second rope from the ground.

INSTALLATIONS

The concrete for a tower is about the same as for a satellite dish pole. (See Chapter 10.) Concrete should extend at least eight inches further than the tower base radius, and generally will be four feet deep. This is sufficient for most TV-FM, Ham, or CB home installations up to 70 feet in height. Sandy soil will require deeper support. Figure 15-2 illustrates the standard installation.

Rohn company and others supply a hinged base-mount that allows you to assemble the tower, and mount the antenna parts, then raise the tower up into position, without ever leaving the ground. Other types of towers generally require installing the base section, climbing up, raising the second section, bolting it onto the first, then raising the third, and so on. Aluminum hoists are available, as they are for satellite dishes, to allow you to raise a heavy section while you guide it into place for the connecting leg bolts. The General Instruments QDMX towers are raised in this manner, although some installers just hoist them up with a rope, and wrestle them onto the top into place.

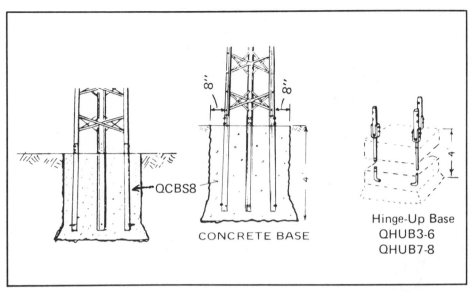

Fig. 15-2. Tower mounting methods in cement.

A less complicated tower is the G.I. QDME (see Fig. 15-1), it can be bracketed to the gable-end of a building for support, then secured to the ground with a base plate held in place with four foot steel base stubs or stakes. If more than 20 feet of tower is above the roof or bracket, guy wires must be used.

Another type is the fold-over model. It is hinged at the center, and built with its own pulley to allow the top half to be lowered, so that it can be serviced without climbing. The cost is much higher than other types, but there can be instances where you will want to sell or service this type of antenna.

Pop-up masts are another method of quickly, and easily, getting an antenna to the required height. The telescoping pole can be embedded in soil or concrete, then clamped onto a structure, or guyed. Since the whole mast (they come in sizes to 50 foot) telescopes, the antenna parts can be mounted, then pushed up into the final position as the wires are dressed and guys attached. Once each section is telescoped into position, a pin is inserted in it to lock it there. Figure 15-3 illustrates this kind of tower.

Since tower work is somewhat of a construction project, one must be totally sure that the job is done correctly. Trainees will be content to lag-bolt a tower house-bracket into a piece of half-inch trim board. On a nice, warm, calm day, it looks sufficiently secure. You will want to assure that support brackets are bolted to solid framing studs. Alternately bolt support brackets to a four-foot piece of angle-iron that spans several frame members. It doesn't take long for you to become an expert at making sure your towers, satellite mounts, and antenna roof mounts are secure, but you have to do it right the first time. A fallen 40-foot tower can create a problem for you, especially if it is now resting on a neighbor's house, or what was their new, $20,000 automobile. High wind is no excuse—you have to be prepared for it. Rohn towers have specifications giving the recommended heights and wind load of the antenna for each size, with wind gusts to 70 mph. Become familiar with the tower brands you handle, and know the security needed for each installation. Manufacturers catalogs have specification sheets for this purpose.

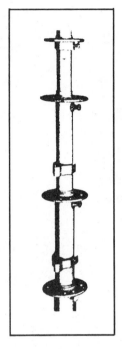

Fig. 15-3. Telescoping "pop-up" tower.

BUILDING CODES

Most rural areas require no construction permits for towers. Some urban areas do. Some municipalities require permits for satellite dishes, and roof-top antennas. Usually a permit can be obtained by submitting an application, and a sketch of the proposed structure you are planning. Tower and roof-mounted satellite dishes, and antennas, can be eyesores, or they can be visually pleasing examples of our hi-tech style of living. The dealer/technician can have a major influence on the aesthetics, as well as the long-term appearance by suggestions and selection at the time of installation. If the local building code for SAM equipment is discriminatory, you and your electronics organizations should attempt to effect a change in it. Such codes or ordinances are usually the result of inaction by electronics dealers in the first place. A little civic effort spent at the time of enactment is the best time to enlighten town officials about the country's attitudes toward free access to electronics communications.

Because towers are frequently the tallest structure in the area, they should always be properly grounded, that means #8 wire link from the tower to an eight-foot copper ground rod. It also means installing a coax shield grounding block on your signal coax at the tower base.

QUIZ

1. A special license is required in most states to perform tower installations and repair work.

   a. ( ) true
   b. ( ) false

2. Working on a 20 foot tower is usually safer and easier than working on a roof-mounted antenna where the roof has a 45° pitch.

    a. ( ) true
    b. ( ) false

3. While instances of antenna workers being injured by a falling tower are virtually unknown, to add an additional margin of safety, a common practice is . . .

    a. ( ) welding additional bracing every 6 feet
    b. ( ) building the cement base an additional 2 feet higher
    c. ( ) installing temporary rope or guy wires part-way up the tower
    d. ( ) using a "cherry picker" crane

4. In place of a lineman's belt, tower work can be done about as well by temporarily hooking your trouser belt around a tower rung.

    a. ( ) true
    b. ( ) false

5. The most practical shoe type to wear when climbing towers is:

    a. ( ) tennis shoes
    b. ( ) thick leather soled shoes
    c. ( ) cowboy boots
    d. ( ) construction work boots

6. A 40 foot tri-leg, self-supporting tower should be planted in concrete how deep and wide?

    a. ( ) 4 feet deep, 2.5 feet wide
    b. ( ) 6 feet deep, 2 feet wide
    c. ( ) 2 feet deep, 2 feet wide

7. Even though a tower can be bracketed to a house gable, the base must be mounted in cement.

    a. ( ) true
    b. ( ) false

8. Since the metal tower is already deeply embedded in the ground, and cement, separate ground rods are unnecessary.

    a. ( ) true
    b. ( ) false

9. All towers need not be climbed. Some are made to fold over near the middle, some are hinged at the ground, and some telescope up and down.

    a. ( ) true
    b. ( ) false

10. Most uhf/vhf antennas are hoisted to the tower top, then assembled.

    a. ( ) true
    b. ( ) false

Chapter 16

Safety Practices

In Ohio, the premium cost for antenna workers' workman's-compensation accident and disability insurance is a full 10 percent of gross wages! Few occupations are charged a higher premium. It might be noted here that higher cost should be considered in your SAM charges. Obviously, state insurance funds must have experience records to justify the insurance rates. The rates should be convincing proof that SAM work deserves extra special precautions to assure that workers are not injured.

THE HAZARDS OF SAM

What kinds of injuries can be expected in working on SAM equipment? Here are some:

- Auto accidents—SAM work requires more travel miles than office and factory jobs.
- Strains—dishes, towers, and antennas can be the cause of hernias. Moving TVs and furniture to accomplish a SAM job is heavy work at times.
- Bruises—a ladder can fall on an employee and cause severe bruises, cuts, or punctures.
- Broken bones—slipping off a pick-up truck bed, jumping off a low roof, falling from a short stepladder, or being hit by a dish, tower, or piece of equipment, can break bones. Trainees especially, often are not used to handling the SAM hardware, and therefore are prone to injuries.

A fellow worker and I were trying to buy a 150-pound ladder-rack for our van. The owner suggested we tie the rack under his building's rafters, and drive our van under it to see if it fit. We tied the front corners up using the clothesline rope the owner assured us was strong enough. We started to tie up the back. The front ropes broke, the rack came crashing down. The truck wasn't under it, and neither were any of us—that time. A 150-pound steel frame falling from eight feet up can be lethal. I was at fault; I didn't demand a stronger rope, instead I trusted someone else's poor judgment.

- Pokes—antenna poles, ground rods, and buttonhook LNA support tubes are sharp objects. So are ladders. Just about everyone has raised up under a pole or ladder that was sticking out over the edge of a pickup truck. Usually this is merely an aggravation. Sometimes it pokes a hole in the skin or causes an eye injury.
- Burns and cuts—soldering irons will burn you. So will drill bits that get too hot, drilling long holes in wood. Cuts occur from poorly made dishes, hack-sawed pole cuts, and sharp edges on tower braces and legs. Most tools can be dangerous. Difficult tower or roof access might require rope safety-anchors. Ropes can burn bodies, especially hands.
- Weather related—snow and ice are hazards on the ground. Handling SAM equipment in a snow, ice, rain, and wind environment adds to the danger. Rain and cold can be the cause of colds and other sickness.
- Other hazards—dog bites and insect stings can be special hazards for SAM people. Cramped attics and crawl spaces have their own hazards, such as smashing a live light bulb with your shoulders, falling through drywall ceilings, getting overheated, falling from makeshift stools and ladders in closet-attic entrances. Portable welders, trenchers, jack-hammers, and powered post-hole diggers can be dangerous. Aluminum ladders can twist and crumple. Wood ladders are often unsafe. Heavy poles and towers can be dropped on your foot. Wet cement can cause serious chemical burns to skin, and especially feet.

Electrical shock is another possibility. Drilling through walls can connect you directly with house wiring. Frequently, SAM workers are wet, lying in damp crawl spaces of basements, or even standing in a muddy trench. Workers have been pushed off ladders, by customers eager to learn how to move their dishes or rotors, before the workers have gotten off the ladders or the roof. Since SAM work requires a large complement of tools, pockets often have sharp screwdrivers, standoffs, eyehooks, or wood screws in them. Sometimes these puncture skin, given the wrong move by a technician.

After reading this listing you can see that SAM work is an occupation that requires safety vigilance. All new workers should be trained and cautioned. Each new procedure, such as hoisting a dish onto the pin, raising a tower, mounting an antenna, securing ladders to trucks, or wearing tool belts, needs explanation and good example. Taking chances on towers, ladders, and roofs is a no-win gamble.

ELECTRICAL SAFETY

Helpers are often not technicians. They aren't electricians. They are sometimes right out of high school. They aren't knowledgeable about shock hazards. Fortunately most SAM equipment is low-voltage operated. Dish actuators operate on 36 Vdc, receivers use 18 volts or 5 volts on terminal strips. Antenna amplifiers are 14 volts ac, and rotors too are low-voltage operated. Most SAM equipment is line-isolated as are TVs. The major shock hazards are from 110-volt equipment, like drills, saws, welders, and house-wiring. There is a small, but not uncommon, possibility that some of the customer's equipment, or the house wiring, is defective. A "tingle" you experience, when touching any metal object in a crawl space or damp basement, is a warning to get your ac voltmeter and see what the problem is. If in doubt, unplug devices plugged into 110-volt outlets. Be careful that your own extension cord isn't the hazard. Check frayed ends and be careful in damp grass, dew, rain, and mud. Don't use portable drills, etc., unless the 3rd wire ground is connected. Don't defeat polarity plugs and adapters. Any power company wires are cause for special concern and precautions. Thousands of people have been electrocuted by power lines. They were people just like you and me.

TRUCK SAFETY

Taking time to instruct employees on truck safety is important. Nearly everyone needs to be shown how to store a 30-foot ladder on a van or pickup properly. If a ladder extends over the rear of the bed, hang a red flag on it. A flag is required on any extension over five feet. Be especially careful of antenna poles sticking over the edge, they are approximately head height.

The numerous hardware items used in SAM work means that some have no established location in the truck. These then are often temporarily laid on top of everything else. From dishes to paper boxes, at 60 mph they will often become hazards to other drivers. Secure everything. Use a tarp over pickups. Tie down the old broken antenna you are bringing to the dump. Don't let nails and screws strew out of the bed onto the road. Keep everything in repair on the truck itself. Build secure racks in vans for tools, the test-TV, the FSM, saws, wire, levels, and so forth. Not only does this secure them, it also lets you find them faster.

ROOF SAFETY

Good ladders are a must. SAM work usually requires extension ladders, plus small and large stepladders. The temptation is always there to overextend the ladder's intended use. Most of the time, it works. Most of the time you won't be injured in a fall.

Some roofs are too steep to walk on safely, even when wearing the proper rubber-soled gripper shoes. Some roof workers use spiked soles for added safety. Steeper roofs may require scaffolding. Lumberyards sell roof scaffold braces that nail under shingles, and will accommodate horizontal boards. Ladder hooks are sold that allow a ladder to be hooked over the peak or ridge of the roof, then climbed down toward the edge. A ladder can be tied to a chimney or some other secure structure on the ground to allow secure roof climbing. While working on a chimney mount, a safety rope can be tied around the chimney and you, in case you slip. The problem isn't that you can't work and walk around the roof work area, it is that certain operations like mounting a dish or antenna pole, can weigh you down, and put you off balance. Old roofs can even have brittle shingles that break, leaving you on a makeshift sled. Snow and rain make special hazards. Wind is always a problem. Antenna rods can stick you in the eye. Poles can knock you off-balance. Bricks breaking off chimneys can surprise you.

CUSTOMER SAFETY

The examples are concerns for you and your crews. In addition, you can create hazards for your customers. Most customers do not make preparations for your work.

Once we installed a coax cable run in a trashy basement that had tools, buckets, wires, shoes, old chairs, and other debris all over. Fuel lines, both in-use, and old discarded ones, lay around on the concrete floor. After our crew left, the customer called—most irate—saying we had broken her fuel oil line. "The basement was covered with fuel oil. What were we going to do about it?" On another occasion, a storm door with a worn-out storm-king cylinder crashed into one of our worker's tool belts, breaking the plate glass panel. Our fault. A chimney mount wasn't installed properly, and it fell off. The antenna crashed to the roof. No harm was done, but it could have crossed wires in another location, or blown onto a neighbor's house or car. While lightning isn't a common hazard for SAM workers, there will come the day when one of your customers has severe lightning damage—perhaps fire. When they are looking for someone to blame, your grounding techniques will come under critical observation. Carry liability insurance.

Other problems, not related to personal safety, can cost you money. Trainees will invariably put the ladder against easily dented aluminum siding. Train them to cushion the rails so as to not cause a dent. Care must be taken with gutters and other parts of old houses. Rugs are a large expense. SAM installers will have to drill holes through them. Always get permission for the exact spot, even if it is the only viable spot. Be sure to seal all concrete wire-entry holes and other entrances, and if appropriate, leave drip loops. Don't leave old broken antennas, and other hardware you threw to the ground in the yard. They can be a hazard. Clean up before you leave. Teach your crews to utilize all the extra trips back to the truck for tools and parts by carrying a handful of junk each trip. When the time comes to leave, everything will be in shipshape.

No one can point out each potential safety problem. The best thing you can do is to instill a safety and efficiency attitude in your crews. If they are safety conscious and attempt to work efficiently, your safety problems will be minimal.

QUIZ

1. Compared with clerical workers, the cost of worker's compensation insurance for SAM workers is at least _____ times higher.

   a. ( ) 2
   b. ( ) 5
   c. ( ) 10

2. Most step-ladders are built ruggedly enough to allow standing on the top or last rung.

   a. ( ) true
   b. ( ) false

3. Since dollies are not commonly used to move TV sets, a TV set is really one of the heavier and more difficult to carry items any worker is usually called on to lift.

   a. ( ) true
   b. ( ) false

4. Slipping off a pickup truck bed and falling is rarely considered a hazard because it is only a three- or four-foot fall. Broken wrists, ankles and wrists can be the result.

   a. ( ) true
   b. ( ) false

5. Eye injuries, puncture wounds, and bruises are more likely in SAM work because:

   a. ( ) poles, ground rods, and antenna elements are sharp objects
   b. ( ) wire ends often snap back toward the face
   c. ( ) radiation from C-Band satellites is concentrated on dish installers
   d. ( ) smoke from chimneys is full of acrid particles

6. Wasps and hornets, frequent residents of antenna and satellite equipment, are a bother but are not considered hazardous.

   a. ( ) true
   b. ( ) false

7. Special shock hazards are most likely in:

   a. ( ) attics
   b. ( ) crawl spaces and basements
   c. ( ) bedrooms
   d. ( ) the shop

8. Getting a "tingle" or small electrical charge, in a damp basement is no cause for alarm, because a tingle is usually ac voltage less the 14 Vac, or only a static charge.

   a. ( ) true
   b. ( ) false

9. A fall on the roof of a house is most likely to be caused by:

   a. ( ) the extreme pitch of the roof
   b. ( ) a sudden gust of wind, a broken shingle, a loose brick, or a swinging antenna pole

10. If you leave a pile of broken antenna, mast, and wire in the yard for the customer to dispose of, and a neighbor child crashes into it and pokes an eye out, you are not liable.

    a. ( ) true
    b. ( ) false

Chapter 17
Solving Difficult Reception Problems

U nusual problems with uhf/vhf TV reception are common. Rather than feeling you are unlucky when you encounter a less- than satisfactory result from your antenna installation, realize all areas have problem spots, not just yours. With satellite systems, it is T.I. or trees. With broadcast antennas it can be many different causes. Some of these are rather difficult to actually recognize. Here is a brief rundown of possible antenna reception problems:

GHOSTS, caused by:

- radio towers
- water towers
- large buildings
- mountains, lake, and creek banks
- power lines
- clouds
- cable company equipment

Figure 17-1 shows a ghost image caused by multiple path signals.

WEAK SIGNALS, caused by:

- trees
- hills

Fig. 17-1. Ghosts on the screen.

- distance from broadcast towers
- wire losses
- splitter losses
- defective TV tuner

INTERFERENCE, caused by:

- power lines, and other electrical equipment
- unstable preamplifier circuitry, or TV tuners
- inadequate antennas for signal conditions
- locally generated interference
- co-channel, and adjacent-channels
- radio stations
- CB, and other communications radios

Figure 17-2 is an example of adjacent channel interference.

That's a lot of possible problems. There are more. Checking over the list, you can probably eliminate many for most sites. You might be in an excellent signal-strength area, with no hills or buildings nearby, far from any radio stations, or cable TV. Most locations are good. Your job is to recognize problems and cope with them. The problem here is

Fig. 17-2. Photo of adjacent channel interference pattern.

that many SAM workers do not recognize a substandard signal. They tend to pass off real problems as bad-reception-areas.

An example: My firm installed a complete new system, antenna, rotor, coax, rotor wire, on an existing 40 foot tower. It was 50 miles from the broadcast stations. Some locations within the nearby 2-mile radius were weak signal areas, and others were good, depending on whether there were trees or hills in the direction of the TV stations. After hooking everything up to the existing in-house coax, the picture was snowy on uhf channels and nearly perfect on most vhf stations. The trainee announced that it "Worked OK, except the uhf channels were a little snowy."

Actually, all twelve stations were substandard. The vhf was better than the customer had been experiencing prior to our work, but most vhf channels were less than 0 dB. I touched the preamplifier power supply. It was overheating, indicating a short in the output coaxial lead. With a short, the 14 volts ac was missing to the preamp. The preamp wasn't working at all. The only signal to the TV set, was what was capacitively coupled through the dead preamp circuitry. Since the antenna was 45 feet high, the signal was sufficient to produce a watchable quality picture on the stronger stations. By checking the impedance across the "To Preamp" coax lead, we traced the problem to an unknown 4-way splitter in a false ceiling. It not only caused the short, and could have burned out the preamp power-supply transformer, but would have decoupled the 14 volts ac to the preamp anyway, if it weren't passive to low frequency ac. The point is: the trainee was willing to accept a poor quality picture rather than attempt to measure the signal or check out the system to see if it was indeed operating correctly. The scenario has occurred with every trainee I have worked with over the years. Unless the trainee is eager to learn the business, they will continue to pass off problems of this type, rather than recognizing and solving them. It behooves a conscientious reception specialist to follow up on every job so that he is sure the employees are capable of recognizing and correcting problems until the system attains its peak potential.

Part of the problem psychologically in the malfunction is usually not evident until all the work is done. The work can be difficult, and take longer than planned. It can involve some unpleasant conditions (dogs, kids, dirty house, trash, cold, or heat). You want nothing more than to see a perfect picture come on when you plug in the amplifier. One of life's disappointments is plugging in the amplifier and seeing no change in the picture. It is about then that the customer informs you, "It's no better than it was!" (Which you are painfully aware of.) So it takes great patience and determination to change your attitude at that point, and begin to debug, or troubleshoot the system. The customer is not sure he called the right expert. Your employees are disgusted and mad. You are late. It's at that point that you are the only person who can solve the problem. It might require climbing back up to the tower, to replace the amplifier. It could be the wires need to be hooked up correctly. A better crimp on the F-connector might be the problem. It might require proving the customer's facts are wrong about the wiring of his system. You are the expert, and must solve the problem. When you do, everyone will be happy.

GHOSTS

Ghosts aren't always caused by reflected signals, but usually they are. In our county we have water towers, a few three- or four-story buildings, a lot of high power lines, a lot of 100 and 200 foot hills, several cement silos, and a lot of metal farm buildings. The seven causes of ghosts listed previously, are not usually difficult to define. The customer is often aware of them. Sometimes he is aware of them, but informs you that he didn't have them on his old system. Why are they there now?

A problem with ghosts is that some poor quality antennas can handle them better than your much-higher-priced model. An example would be when signals come from multiple broadcasting directions, as shown in Fig. 17-3. If your new antenna has a good forward polar pattern it will tend to receive the station B signal poorly from the rear, but readily accept the station B reflection from the building at C.

On the other hand, a conical antenna has a polar pattern that is poor at rejecting rearward signals. Therefore, the desired station direct-signal from B is received cleanly, and is strong enough to swamp out the weaker, reflected-signal off building C. In Fig. 17-3, the more directive antenna rejects the rearward signal and accepts the reflection, causing an objectionable ghost. This could well be a case where your recommended antenna works worse than the 20-year-old broken conical!

A better solution of course is to install a rotor and use an even more directional antenna than in Fig. 17-3. The more directional, the better able it is to reject ghosts at smaller side angles, and to accept the desired signal. If you are not successful in doing it right (with a rotor), you might want to experiment with several antennas, until you get the best compromise. Figure 17-4 demonstrates the advantage of a conical antenna.

The other causes of ghosts can be usually located, using the calculations in Table 17-1. Bear in mind two things: A TV antenna will rarely pick up reflections that are coming from its side quadrants; reflections that travel to an object rearward to your antenna, have to travel past the antenna to the object, then travel back to the antenna. The signal then takes twice as long, at this distance and direction, meaning the rear reflective object is half as far away, as it measures from forward directions. Figure 17-5 demonstrates how power lines cause ghost images.

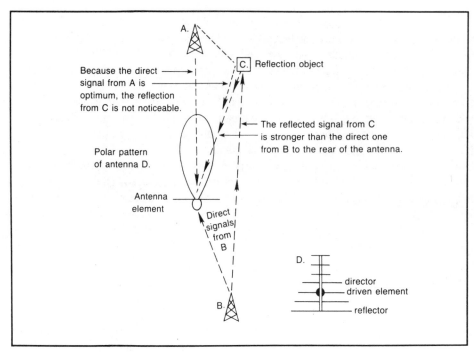

Fig. 17-3. Drawing of reflected-signal paths.

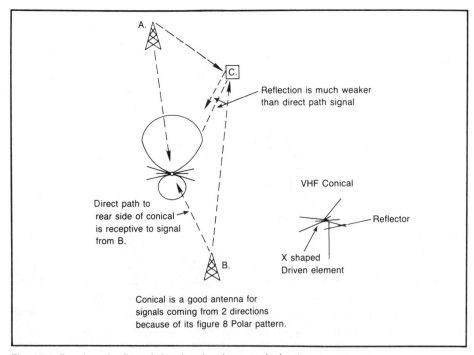

Fig. 17-4. Drawing of reflected-signal path using a conical antenna.

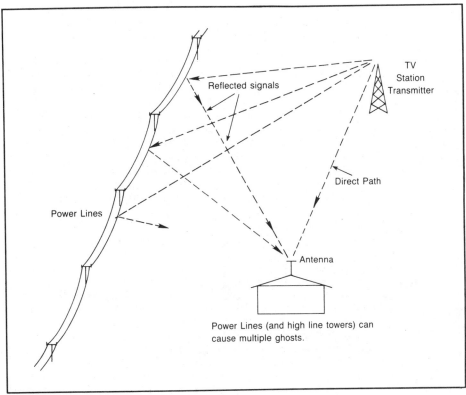

Fig. 17-5. Drawing of reflections from power lines.

WEAK SIGNALS

There are three primary ways to improve weak signals:

- Use a higher gain antenna
- Use an amplifier
- Raise the height of the antenna

A fourth method is to improve the system by replacing vhf-only components with 900 MHz parts, replacing high-resistance wire, and removing actual signal-reducing problems, like improper lead-dress.

Judging which route to take is something one gets from experience. Fortunately, each possible solution could help, and unfortunately, all might be required.

To solve the problem with a higher gain antenna, you need to realize that a 50 element, long-range antenna will improve signal strength by less than 12 dB over a short-range antenna with less than half the number of elements. If you require 20 dB more for satisfactory signals, a larger antenna won't completely solve the problem.

If your signal is −40 dB, which is just about no-signal, a 17 dB amplifier won't do the job. If the signal is too small, amplifiers will amplify the snow. Your signal will still be unsatisfactory, even though your field strength meter might read a surprisingly high level that would ordinarily produce a good picture.

Table 17-1. Ghost Calculation Table.

*1 000 000.	1 million
* 100 00.	100 thousand
* 1 000.	1 thousand
* 1.	1
* .1	1 tenth (100 milli)
* .01	1 hundredth (10 milli)
* .001	1 thousandth (1 milli)
* .0001	1 ten thousandth (100 micro)
* .00001	1 hundred thousandth (10 micro)
* .000001	1 millionth (1 micro)

Speed of light (and radio waves):

* 186 284 miles per second
* 186.284 miles per millisecond
* .186284 miles per microsecond
* .983579520 feet per second
* 983579.520 feet per millisecond
* 983.579520 feet per microsecond

Television screen sizes (diag. & actual):

5.5	7.0	10.0	12.00	13.0	19.0	20.0	25.0
4.5	5.5	8.5	9.50	11.0	16.0	16.0	21.0

Scan time of TV is 63 usec.
Determine ghost percentage of full screen.
Take that percent of 63. Gives the time of delay causing ghost.
Take that time times 983.57952 (distance in feet signal travels in one usec.)
Half of this is the distance to the reflective object in feet.
Convert this to miles or blocks or meters etc. as desired.

Screen widths in fractions:

Diag.	Act.	16ths	8ths	4ths	1/2's
5.5	4.5	72	36	18	9
7.0	5.5	88	44	22	11
10.0	8.5	136	68	34	17
12.0	9.5	152	76	38	19
13.0	11.0	176	88	44	22
19.0	16.0	256	128	64	32
20.0	16.0	256	128	64	32
25.0	21.0	336	168	84	42

Graph allowing you to judge the angle of a reflector. Bear in mind that reflections near 90° from the TX-REC path are less likely to be a problem than rear ghost or those from 45° due to the polar pattern of most antennas.

Screen sizes

GHOSTS	25	19	13	5.5"
G				
H	1/4"	3/16"	1/8"	1/16"
O	3/8	1/2	3/16	1/8
S	1 1/2	5/8	5/8	1/4
T	2	1 1/2	1 1/4	1/2
S	2 1/2	2	1 1/2	5/8
	3	2 1/2	1 5/8	3/4

Distance from primary object

Infinite angular reflection points

Transmitter

Miles: .1 .2 .3 .4 .5 .6 .7 .8 .9 .10

1μS, 2μS, 4μS, 6μS, 8μS, 10μS

Receiver

Feet: 500, 1000, 2000, 3000, 4000, 5000, 6000, 7000, 8000, 9000, 10000

Tables for rear reflectors causing ghost

*** 5.5 in. screen (4.5 act.) ***

Ghost size (in.)	Screen delay (usec.)	Reflector distance (ft.)	Reflector distance (mi.)
1/8	1.75	861	2/10
3/16	2.63	1293	2/10
1/4	3.30	1721	3/10
3/8	5.25	2582	1/2
1/2	7.00	3443	6/10
3/4	10.50	5164	1
1	14.00	6885	1 3/10
1 1/2	21.00	10328	2
2	28.00	13770	2 1/2
3	42.00	20655	4

*** 13.0 in. screen (11.0 act.) ***

Ghost size (in.)	Screen delay (usec.)	Reflector distance (ft.)	Reflector distance (mi.)
1/4	1.43	703	1/10
3/8	2.15	1057	2/10
1/2	2.86	1407	1/4
3/4	4.30	2115	4/10
1	5.73	2818	1/2
1 1/2	8.59	4224	3/4
2	11.45	5631	1
3	17.18	8449	1 1/2
4 1/2	25.66	12675	2 1/2
6 1/2	37.23	14308	3 1/2

*** 19.0 & 20.0 screen (16.0 act.) ***

Ghost size (in.)	Screen delay (usec.)	Reflector distance (ft.)	Reflector distance (mi.)
1/4	.98	482	1/10
1/2	1.97	482	2/10
3/4	2.95	1451	1/4
1	3.94	1938	1/3
1 1/2	5.91	2906	1/2
2	7.88	3875	3/4
3	11.81	5808	1
4 1/2	17.72	8714	1 2/3
6 1/2	25.59	12587	2 1/2
9 1/2	37.41	18396	3 1/2

*** 25.0 in. screen (21.0 act.) ***

Ghost size (in.)	Screen delay (usec.)	Reflector distance (ft.)	Reflector distance (mi.)
1/2	1.50	738	1/10
1	3.00	1475	1/4
2	6.00	2951	1/2
3	10.89	5356	1
4 1/2	13.50	6639	1 1/4
6	18.00	8852	1 2/3
7 1/2	22.50	11065	2
9	27.00	13278	2 1/2
10	30.00	14754	2 3/4
12 1/2	37.50	18442	3 1/2

Using RG-6 coax instead of RG-59, can give you a slight improvement. Using the highest gain preamplifier (some are 30 dB) will help. Obviously, using the highest-gain antenna is a plus. If you are starved for signal, work on optimizing it to only one TV at first. After it is satisfactory connect to the splitters and taps, if used. If you can get the signal to one TV, you can install a line amplifier to compensate for multiple sets and long coax runs. If you can raise the height of the antenna even four or five feet, it can help. Occasionally, one end of the house will give a slightly higher signal. Use your signal strength meter and test antenna to see if a better location is possible.

Below a certain level of signal, around −40 dB, the only solution to a weak signal is to get higher above the terrain. If your location is immediately past a 200-foot hill, and you change the height from 25 feet to 50 feet, it might not help. You might need an 80 foot tower (or taller) to obtain a signal sufficient to override the ambient terrestrial, amplifier, and TV mixer circuit noise. To keep out of trouble in these locations, carefully check your area topography maps, and don't commit to guarantee the results without entering into an agreement to be paid for your engineering, and experimental time.

Besides an rf meter, carry a test TV set, one that you are familiar with. Often a bad TV is the reason you are doing the antenna job. Proving the system is working can save you much troubleshooting time.

CABLE LEAKS

Leaking cable TV systems can be difficult problems to diagnose. Especially if the cable channels are the same as the broadcast signals. Some cable TV drops using RG-59, are 500 feet long. The signal in that coax can be +30 dB or more. If the F-connector at the pole or trunk line is loose, you can have a healthy sized rf signal being transmitted all over the neighborhood. This signal will be delayed by 2 to 10 microseconds (or more). This competing and delayed signal, if stronger than your directly received broadcast signal, will cause the TV to try to lock on both sets of sync pulses. It can't, so the picture weaves and fades in and out. It very likely will be totally unwatchable. You will assume the problem is caused by a reflected signal.

Once you suspect the cable, walk around the area with a directional test antenna. Radiating cables are not easily detected, but once you do find one, you will soon recognize the symptom. Since it is illegal for cable systems to radiate a signal −40 dB or greater, at 10 feet from their equipment, they are usually willing to make a quick repair. Frequently the problem isn't a bad F-connector, but one or more illegal tap-offs, made by subscribers. College fraternities, and apartment houses are notorious for having multiple illegal splits. College students are rarely good at choosing the proper wire, or at making good connections. Disconnecting an entire building might be the only way you can locate the problem.

POWER LINES

Power line sparklies are difficult to diagnose. The source of the interference can be several hundred feet away. It can be an arcing pole transformer or capacitor. It can be a bad insulator on a high line. Electrical equipment can cause baffling problems. An arcing start-up capacitor in a compressor, a machine shop arc-boring tool, microwave ovens, cash registers, computers, and adding machines, hog-house heaters, even some light bulbs can cause interference. Electronic, ultrasonic, anti-bug and vermin devices, and wireless phones, are common causes of objectionable interference patterns. Each source usually has its own distinctive pattern. Once you identify a source, you can deal with it. Obviously, the stronger the available signal, the less effect the objectionable interference will have on it. Your odds of eradicating power line sparklies, 60 miles from a broadcast station, on a TV set using rabbit ears for an antenna, are not good!

PREAMPLIFIER PROBLEMS

Rarely will you install a preamplifier that doesn't work. They are usually very reliable. You might find one that causes your TV set to receive stations on one or more harmonic locations. Channel 20 might come in on 20, 40, 50, 71, etc.

I had a preamp that caused only one of the available six channels to be missing. Another produced excessive herringbone patterns. Preamps might also have reduced output. They are subject to lightning damage.

TV SETS

In addition to weak tuners in TV sets, sometimes the tuner is unstable. A perfect signal, now stronger than the customer ever had, could cause the tuner to go into spontaneous oscillation. Herringbone patterns, squeals, and blanked out video and audio can result. While you might be able to readjust the rf and *if* AGC control settings to solve the problem, that might be a chance you don't want to take. If your test TV works perfectly, show it to the set-owner, then discuss the needed TV repair job, and the price. If the customer is suspicious that your antenna is the real cause of the problem, and the TV was just fine before you came along, you might be able to solve the problem by inserting a variable attenuator in the signal line, and adjusting the signal level to a point high enough to produce an adequate picture, yet low enough to reduce or sidestep the unstable tuner problem.

CO-CHANNEL AND ADJACENT CHANNEL INTERFERENCE

You are the hero when you produce one or more additional TV stations for your customer. In some rural areas you might find the available channels are not separated by a blank channel. For instance, you could have a channel 2 and 3, or a 12 and 13 available. If one is considerably stronger than the other, it will frequently overlap, causing adjacent channel interference. If a rotor and directional antenna can't solve the problem, the answer becomes more expensive. Homeowners might not want to pay the price. Apartment house owners, and MATV system operators often eagerly accept a fix that gives them more choices. One common method of solving adjacent channel problems is to use two single-channel antennas, inputting each signal into a separate selective band-pass filter. The filter chops off the unwanted channel signal with very little desired-channel attenuation. One of the channels then can be converted to some other blank channel location, perhaps uhf, then re-introduced into the system. The strong (10 dB) TVRO receiver output normally interferes with off-air channels 2, 3, or 4. In a single-family dwelling, the TVRO receiver can be turned off to eliminate the problem. Using a video switch-box can be a solution, but the isolation between ports usually is not sufficient to completely eliminate cross-talk-FM herringbone patterns on off-air channels 2, 3, or 4.

Co-channel signals most frequently produce an even pattern of horizontal gray bars about ⅜-inch to ½-inch apart, usually moving slowly up the screen. When severe, the same-frequency signal will completely override both sound, and picture carriers. Especially in June and July, the skip conditions will cause same-channel reception from over 1000 miles away. Low vhf channels are the most frequent victims. There might be no solution during these periods, other than using a highly directional antenna and a rotor, to aim away from the skip signal. If the co-channel is a local channel 3, and your customer has a VCR, video disk, and satellite dish, you might find converting off-air channel 3 to an unused channel is the best choice.

RADIO STATIONS

Radio station interference can be reduced by inserting a trap at its frequency, or its first harmonic. TV channel 8 is at 180-186 MHz. People living near FM broadcast stations using 90 to 93 MHz will be affected on channel 8. Sometimes a simple FM trap is enough to reduce the objectionable interference. Sometimes, more expensive and selective traps with sharper or deeper notches are required. In severe cases the unwanted radio signal can be inverted, and reintroduced into the line, balanced in level, and then canceled.

CB interference can be reduced by using a hi-pass filter that chops off everything below 52 MHz. Since the TV set itself contains such a filter, you are merely doubling the effort. If the CBer or communications radio station is so close, and so powerful, that added filters are required, it is a problem that might better be solved by the interfering party, if he can be identified.

QUIZ

1. A TV ghost is usually located to the _____ of the primary picture tube image.

   a. (　) right
   b. (　) left

2. Using a 19 inch TV screen, you notice a ghost, to the right of the primary picture by 1 inch. The reflective object is thought to be off the antenna boresight by about 30 degrees, and is about ⅓ of a mile away. Could this object be the cause of the 1 inch ghost?

   a. (　) yes
   b. (　) no

3. The most likely cause of a strong ghost, which slowly changes its distance from the primary object on a TV screen, might be:

   a. (　) metal barn
   b. (　) smokestack
   c. (　) cloud
   d. (　) leaky cable TV dropline

4. A -50 dB antenna signal can be improved to a satisfactory level by:

    a. ( ) elevating the antenna
    b. ( ) installing a preamplifier
    c. ( ) installing a line amplifier
    d. ( ) changing to RG-6 cable

5. Signals emanating from cable TV trunk lines or drops can be recognized by:

    a. ( ) forward ghosts in the picture
    b. ( ) trailing ghosts in the picture

6. Two bands of sparkly interference, especially noticeable on low vhf channels can be

    a. ( ) cable TV leakage
    b. ( ) co-channel interference
    c. ( ) adjacent channel interference
    d. ( ) power line interference

7. Reduced output from a preamplifier can be detected by:

    a. ( ) measuring the antenna signal, then measuring the preamp signal output
    b. ( ) it can only be detected by exchanging the preamplifier

8. If the improved, stronger signals from your new antenna system cause the TV set to go into oscillation, or blank out on one or more channels, you should:

    a. ( ) remove the system
    b. ( ) repair the TV
    c. ( ) suggest the TV be repaired, and insert an attenuator to temporarily solve the problem.

9. To solve adjacent-channel interference problems where both channels are desired:

    a. ( ) equalize the signals with bandpass filters
    b. ( ) install a rotor
    c. ( ) add a line amplifier
    d. ( ) install a FM trap

10. To reduce co-channel interference:

    a. ( ) use a highly directional antenna
    b. ( ) use a higher gain amplifier
    c. ( ) use a bandpass filter
    d. ( ) use a bandpass trap

Chapter 18

TVRO Troubleshooting

Dish system troubleshooting might be divided into two categories:

- Installation problems
- Breakdowns in an operating system

Breakdowns occur due to the following reasons:

- Insufficient weatherproofing
- Inadequate product weather resistance
- Wind damage
- Abuse
- Lightning, or other over-voltage spikes
- Cable problems
- Customer-caused problems, such as weak receivers, or VCRs
- Electronic malfunctions

Installation problems are:

- Non-perpendicular poles
- Inadequate base, pole, or mount
- Poor, or non-existence, dish and mount instructions

- Missing dish and mount hardware
- Bad cables, and bad cable connections
- Terrestrial interference
- Trees, wires, or building blockage
- Improper angle, polar bar, and focus presettings
- Defective reflector
- Customer nuisance calls

INSTALLATION

Obviously, doing the installation correctly will prevent some later breakdowns.

The easiest problem to prevent is the non-perpendicular pole. Anyone can use a level. Getting the pole vertical is easy to do. As the 3 foot deep by 2 foot wide hole is filled with concrete, the pole will stand by itself, and can be tilted easily. By the time the hole is completely filled, the pole is tight, and it takes more effort to effect a slight correction. Some poles do not have smooth circumference surfaces. Some are out of round. Some are slightly bent. The object is to make sure the top portion, over which the dish mount sleeve is positioned, is perpendicular. Some trainees make one or two level checks, and pronounce the pole straight. If you accept a less than optimum pole setting, you will pay the price when you attempt to span the satellite arc. You will find yourself attempting to correct it, with elevation and declination resettings. You might never get F1, F5, or F2 as easily as you would, had the pole been straight.

I installed a 20-foot pole on a sidewalk, using a collar at the building roof, then noticed the two feet of pipe at the very top was bent. I hired a local welder to cut the top two feet off, and rewild it straight.

In our type of clay soil, a two foot by three foot hole is more than adequate. This requires about 10 bags of concrete. Some installers use as little as four bags. I've never found a pole installed with so little cement to have twisted or tilted. I suspect four bags is sufficient in hard, dry, packed clay dirt. In wet damp soil, or other areas of the country that have sandy soil, three feet by three feet might be necessary for a 10-foot dish. We have installed poles in dirt alone, at county fairs or other temporary places, and had no problem with twisting or tilting.

Most poles are ¼ inch steel casing. Cement can be added to the inside to strengthen it, but it is not required for a 10- or 12-foot dish. The breakoff point is at ground level. Filling the pipe a foot or more above ground level, even all the way to the top, is insurance against 100 mph winds.

The mount on many dishes should fail before a pole breaks. Thousands of dishes are held onto the four pad attachment legs by only four bolts, either ⅜-inch or ½-inch. The wind has only to gust enough to rip the dish material, and shear these four bolts. Most modern dishes have the mount attached to braces, that provide many times the necessary strength. Some dishes have secure dish anchors, but have weak, angle-iron mount frames, or brackets. These break in high winds. They also break when, or if, the actuator gets in any kind of a bind, or overextends.

Some mount sleeves do not fit the diameter of the pole. Whether it is too small, or too large, it doesn't matter. You will need to weld the next size down to your pole, or slip the next size larger pole-shim over it, before you can mount the dish.

If you have only a small gap between the pin and mount sleeve, compensate for it by making sure the sleeve set-screws are each tightened, so that the gap around the pole is equal at all points on its circumference. Shim material inserted at three points in the gap sometimes solves the problem. This is to make sure the sleeve and dish mount are not tilted, even though the pin is perpendicular.

Bad Mounts

Many mount problems have been caused purely and simply by poor quality. Some manufacturers and distributors have tried to reduce costs by substituting inferior mounts. Invariably the substitution was more difficult to mount, and usually contained no instructions. Installers spent many extra hours trying to figure them out. When you reach a point of rechecking the declination angle and resetting it, you have already reached the dish installer's frustration limit! One of the best answers to that problem is simply using known, good, hardware. However, the nature of the TVRO business is to try new products, perhaps on the advice of fellow dealers, or distributors. Each new dish, and each new electronic product, requires a sort of apprenticeship, before all the peculiarities, and potential traps are recognized. Figure 18-1 shows a slightly malformed satellite dish.

Dish makers have been garage operators in some instances. The low-priced, fiberglass, and spun aluminum dishes were in favor for a while, because the price was lower. The cost-cutting was at the expense of good literature, and quality control. Experienced installers could assemble the parts for practically any dish. Newer technicians would spend hours, trying to make wrong parts work, or to aim or focus a dish that

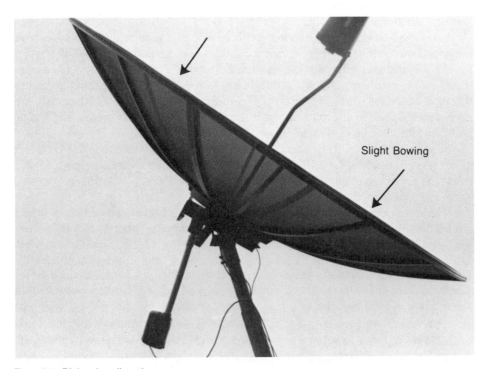

Slight Bowing

Fig. 18-1. Dish edge distortion.

could not be adjusted sufficiently. The lost hours cost hundreds of dealers their businesses. Instructions for many popular makes of dishes were either non-existent, or poorly done. A popular 10 foot mesh dish for example, begins the construction process with the sentence: "Identify the flattened end of the dish ribs." If the flattened end (Is it the end, the concave top face of the rib, or the convex bottom face?) has to be identified, there is something wrong. It should have been painted orange, or a paper sticker attached, or not anodized. Put a dish together with one rib backwards, and you will likely re-assemble it on your own time.

Problems Are Likely

The possibility of installation problems occurring is more likely than the possibility of them not occurring. If the pole is straight, and the right size, and the mount fits perfectly, the focus is correct, the LNA is centered, the declination and elevation are correct, and the polar bar is zeroed in on the North Star, you can still have problems with the N-connectors between the LNA and downconverter. The RG-6 coax cable wire won't fit in most LNB female F-connectors. Feedhorn polarizer servos go bad frequently; receivers and actuator controllers die in front of you. You can spend valuable time troubleshooting VCR, TV, and distribution system problems. Terrestrial interference is always a threat. Sometimes you will find your actuator controller circuitry is the cause of unacceptable herringbone patterns in the video display. A percentage of installations are affected on one or more satellites by trees. While you might have made a pretty good site survey, and casually mentioned that one or two westerly, seldom-watched satellites do not come in clearly, due to the trees on the horizon, the dish owner later might demand to know why NBC comes in snowy on F2! What you know is a fantastic installation, receiving all the other satellites perfectly, might not satisfy the customer, who is trying to get every penny's worth he paid for.

Cable problems are usually caused by the connections at the dish, or in the house. Placing the terminal strip and connections often made at the pole in a weatherproof junction box, is a popular way of reducing the chances of cable problems. Feedhorn servo wires must be spliced securely. The LNA N-connector, if used, should be screwed together gently. F-connectors to undersized LNB connectors, must have barrel splices, and male-to-male F-unions, to insert properly. Forcing a #18 wire in a #22 hole will sometimes work, but usually only for a short time.

TERRESTRIAL INTERFERENCE

T.I. as discussed in Chapter 12, is a special problem. Recognizing it is the first step toward defeating it. Blaming trees, moisture, or aiming problems on T.I. only aggravates the situation. Trees, of course, can't be tolerated. Even the tiniest branches will severely attenuate the satellite signal. If a customer can't remove the offending tree, your job is to explain the viewing penalty clearly. If you expect only one clear boresight shot at a satellite, then that fact must be conveyed to the customer, before you dig. With experience, you will find a clear boresight cylinder can still be affected by trees that are about 10 to 15 degrees of boresight. Don't expect the same quality as you get in an open field. There can be a canceling effect, caused by satellite signals glancing off tree leaves, and moisture in branches, not directly in front of the dish, but slightly off to one side.

ALIGNMENT

Adjusting a dish can be simple. Set the elevation for your latitude. It is 40 degrees in Central Indiana. Set the declination as instructed by the dish maker. Our D.A. is 6.2 degrees. Use a compass, and set the polar bar to true North. We have no magnetic deviation in Western Indiana. Add or subtract for your area, or go out at night and aim at the North Star. If these are set right, and focus is set according to the manufacturer's specified distance, you should turn on the receiver and immediately pick up some blanking bars, or color from a satellite channel. If you do, mark the positioner location if it has a readout, so that if you get nothing else at this point, you can return and start over.

Try to move the dish as close to a due south satellite as possible. We use T2 and F4. If you get a picture, adjust elevation for best picture, or highest meter reading. You will probably use a helper and walkie-talkies to accomplish this. Then run the dish westerly to F3, G1, F1, or F5. If that satellite is also perfect, the North-South polar bar alignment is correct. Usually it isn't. If you can raise the lip of the dish, or lower it, and cause an improvement, then N-S is slightly off. Loosen the pin. If raising the dish helped, then twist the dish slightly on the pin, to the west. Tighten the pin set screws, and return the dish to the satellite. If no further improvement is seen by raising or lowering the dish, the N-S is fine. If lowering the dish causes an improvement, move the dish more easterly.

NO PICTURE

If you have everything set correctly and have no picture, the cables or electronics could be at fault. If you have a solid gray or black screen, chances are the downconverter is not working. Check the 18 volts to it. If it is working you will get the unique satellite snow. Some TV sets and some satellite receivers don't seem to produce the satellite snow signal strongly, and technicians mistake weak satellite snow for a no-snow condition. If you have the proper type of streaked snow, yet no picture, the LNA can be bad. Replace it. If you have replaced the equipment, and still have no picture, make sure your TV channel 2 (or 4) is correct. Some TVs have their fine tuning misadjusted. Newer sets have overcome this possibility. If nothing shows, start over. Check angles, voltages, and focus distance.

Some 70 MHz receivers will work fine, but won't tune the lower or higher channels. This could be caused by a bad downconverter, and can be a frustrating problem. I have had it happen on block systems. Other troubles that can stymie you are: wrong digital readouts, burned out indicator bulbs, excessive herringbone or hash interference generated internally, loose chassis connectors, dead modulators, and poor sound quality. Being able to substitute a known good receiver is a nice solution.

PROGRAMMABLES

Most modern TVRO receivers and actuators are programmable. Each manufacturer inserts his own programming codes. Like home and business computers, you are required to do everything right. Nearly right isn't good enough. While becoming familiar with a new model and brand you might become confused. You can misinterpret the requirements, and conclude the receiver/actuator controller is bad. Operator manuals aren't always helpful, and your particular problem is usually not listed. Start from the beginning, and go through the procedure slowly, step by step. An outstanding example is the Luxor 9550/9534. You must use the remote control. You select a satellite, center the dish position, set the odd polarity position, set the even polarity position, then press STORE. Most other positioners require you to press STORE after both the odd polarity settings. You can try to store odd, then even, for hours on the Luxor; it won't retain the correct position unless you do it according to the book. When programming doesn't seem to respond properly, sit back, read the instructions in detail, and proceed one step at a time, exactly as specified.

Strange Things Are Happening

On rare occasions, after everything is hooked up properly, and all fittings and connections are perfect, you receive only a jumbled picture. Unusual modulation, hash, herringbone and audio problems might exhibit themselves. Sometimes there is a defect in the electronics. Before condemning it, consider that computers sometimes act in a similar manner, locking up, and seeming defective. Unplugging or de-energizing the computer may reset all voltages, correcting the problem. Some gremlin-like, logic-defying, TVRO system problems are similar. Turning the receiver off and on doesn't eliminate the garbage. No amount of channel changing, polarity reversing, or other frantic switching helps. Unplugging the polarizer, with the power on, then plugging it back in, does! No wiring problem. Perhaps the receiver or controller doesn't have a quick enough reset ability, and a combination of wrong sequences sets up an unwanted oscillation. The object here is to use all your logical troubleshooting techniques, making sure no loose or shorted wires are the cause of the problem, and that the receiver functions are correct. If all this fails, start over from an unplugged condition. With the power on, disconnect the peripherals, the LNA/DC lines, polarizer, and actuator/drive. Plug them back in, with the power on. If the problem persists, trade electronics, one part at a time, until the trouble is cleared. We have had the above symptoms on several brands after thunderstorms.

WEATHER-RELATED PROBLEMS

Problems with previously-operating systems frequently are caused by the weather. Most early actuators did not have good O-rings and seldom were mounted with bellows-type weather covers, and motor boots. Water seeps inside, it ices up, or causes rust. Water can fill up the sensor and gear compartment, if the weep-hole is not installed downward. A very common problem with actuator arms is improper installation. Since the linear actuator must pivot on the mount, a plastic separator washer must be used. A metal washer will freeze or rust. At the reflector bracket, the telescoping arm must clear the dish brace at both ends of the satellite arc. If the entire actuator is binding, or at some crazy angle, take it off, or re-engineer it, so that it is not binding. Seal all cable connections, and places water might ever enter. Figure 18-2 shows an actuator that needs more clearance.

FEEDHORN POSITIONING

LNAs must be set at the exact focal point of the dish. Assuming you can merely adjust the focus at the "best picture" spot will cost you time. Set it first at the exact

Fig. 18-2. Check the actuator arm assembly for sufficient clearance.

distance specified from the reflector plate. Try to figure out whether the instructions mean the distance from the plate to the polarizer opening, or ¼-inch inside it. After you have a good picture, then and only then attempt to improve the picture by moving the focus ½-inch in or out at a time. Obviously a sensitive signal strength meter is desirable to observe small improvements. Pay attention to the polarizer instructions for setting the scaler rings. If non-adjustable, use a gold-ring on deep dishes, and reset the focus. Always make sure the LNA is centered on the dish. Gently push up and sideways, while your co-worker monitors the picture for improvements. You might need to shim the LNA support tube, or even guy wire it into the center position. If the LNA has no weather cover it will eventually get water in the waveguide elbow. Dry it out and install a feedhorn cover. For your own sake, try to install covers with very small air holes. If ¼-inch or larger holes are in it, you will have to fight wasps later.

ACTUATORS

Actuators are relative simple to troubleshoot. You can substitute another motor drive in the house, before checking the outdoor unit (ODU). Potentiometer types will usually move, even though the sensor and readouts are inoperative. You can measure the 10K potentiometer with an ohmmeter, observing the changes in resistance. If the potentiometer is good, and the readout isn't, chances are the display circuitry or counters are bad. Open motors have bad brushes, or brushes that can't press against the commutator, because moisture and slime have glued them in a fixed position. Some motors get hot and burn out, if the shaft is bent and rusted. See if you have 36 volts dc going out to the motor, and at the motor terminals before dismantling the motor drive. Optical sensors and bad reed switches in actuators are best found by substituting a new motor drive, then troubleshooting the old drive at the shop. Figure 18-3 shows the inside of an automatic actuator.

150

Fig. 18-3. UNIDEN automatic actuator.

Blown Fuses

When you or your crew members short the dc power to ground, by carelessly touching the 18 V or 5 V to the chassis, you might blow a fuse. In some Uniden products (and others), the fuses are inside, in addition to the main fuse, available on the back panel. Getting familiar with your products is the secret to avoiding being stymied by simple problems.

RELATED PROBLEMS

Some problems appear to be related to your equipment, but they are really caused by the TV, VCR, or maze of wires that are common now in many homes. It is aggravating to discover the off-air antenna has developed a problem, just when you thought you had a perfect installation. After hooking up dozens of off-air antennas to satellite receivers, I hooked up one to an Amplica 300 and didn't have any uhf channels. I assumed the internal a-b switch was defective. After an expensive change of equipment, and shipping for repair, the truth came out . . . the Amplica doesn't pass uhf!

Broken dish-to-house wiring is common. If the connections are OK at the receiver, and no problem is evident at the dish, open each cable at one end and check for shorts with an ohmmeter. If no short is indicated, short the wires together on one end, and check from the other end, for continuity of each. That should prove the integrity of the cables. If open, look for obvious places for breaks. Examine the entrance to the house, wherever gardening has taken place, and the basement, where any activity occurs. Look for possible damage caused by animals.

UNFAMILIAR GEAR

You will have to make service calls on unfamiliar equipment. Take what you have for substitutions: LNAs, LNBs, feeds. Chances are you will not be able to make a repair on site. You might have to bring the actuator, receiver, downconverter, or other part back to the shop, and then to the maker or other service depot. Carrying a spare receiver system that you know can easily be substituted, can eliminate some of the electronics, or identify the exact bad component. With a complete service department you will learn to perform service on an increasing number of brands, as you increase your familiarity with them.

QUIZ

1. The one mounting pole you brought with you to the site has a slight bend about two feet from one end. The pole should be mounted:

    a. ( ) bent end in concrete
    b. ( ) straight end in concrete

2. Since steel tubing used for dish pins is round, a level check at any point will always show whether the pole is actually perpendicular

    a. ( ) true
    b. ( ) false

3. What must be done if the dish sleeve inner diameter is slightly larger than the o.d. of the pin?

    a. ( ) replace the pole
    b. ( ) replace the dish sleeve
    c. ( ) compensate by adding slightly to the declination angle
    d. ( ) insert three or more strips of shim stock in the gap

4. A percentage of female coax chassis mount sockets are too small to accept RG-6 coax. What is the best solution?

    a. ( ) file down the RG-6 center wire
    b. ( ) use RG-59 F-connectors rather than RG-6 F-connectors
    c. ( ) install a barrel splice and RG-59 F-to-F connector
    d. ( ) cut the #18 wire in the RG-6 at a slight angle and push it in

5. The polar bar should be set to

    a. ( ) magnetic North
    b. ( ) true North
    c. ( ) the evening star

6. A frequent TVRO problem causing a completely blank screen (no snow) on the TV, might be:

    a. ( ) no power to the downconverter
    b. ( ) bad polarizer

c. ( ) dead LNA

d. ( ) open N-connector to the LNA

7. You find the actuator will not move the dish. In checking you find the motor and sensor wires were hooked up backwards. After reconnecting them properly the dish will still not move. What might be the problem?

a. ( ) motor is burned out

b. ( ) sensor is burned out

c. ( ) 10 K potentiometer is burned out

d. ( ) control console has to be bad

8. Everything on the TVRO receiver works fine, except that the feedhorn polarity will not adjust. At the dish you find all three polarizer wires have 5 volts on them. What is the problem?

a. ( ) servo controller is shorted

b. ( ) servo motor is locked up at one end stop

c. ( ) ground wire is open

d. ( ) pulse wire is open

9. If a new model programmable receiver/actuator appears not to program properly:

a. ( ) ship back to factory

b. ( ) replace the memory batteries

c. ( ) check for wiring problems at the dish

d. ( ) read and follow instructions again

10. If the dish is properly aligned to track the visible Clarke Belt satellites:

a. ( ) moving the lower lip of the dish up or down slightly will not affect the picture

b. ( ) raising the dish lip should cause the picture to improve slightly when aimed due South only

c. ( ) raising the dish lip should cause the picture to improve slightly when aimed either maximum west, or maximum east, but not due south

d. ( ) moving the dish lip either up or down should worsen the picture quality

Chapter 19

Troubleshooting Antennas
and MATV Systems

A TV set can be utilized as a signal strength meter. After all, the quality picture on a TV is the end result of your work. A rugged portable TV can be a valuable test tool for the antenna technician; however, it is limited. While it will indicate whether the system is producing a snow-free picture, you can't really tell the difference between a 0 dB or a 20 dB signal, nor is it easy to tell whether you have a -20 dB or -10 dB signal. Without the ability to accurately measure the signal levels throughout the antenna system, you must do a lot of guessing. The guessing, cutting, and trying, will cost time. It can cause you to produce a lesser quality job. The only answer then is for you to own an rf signal-strength meter.

THE RF METER

There are several brands of rf meters made: Leader, Sencore, and Channel Master are three. The rf meter should be tunable over the 54 to 900 MHz band. It should be battery operated, and include 50 to 100 dB of optional attenuation, or some other means of measuring levels between -40 dB and $+80$ dB. It should have an automatic shut-off, either timed, or cover-operated. It should have adapters to connect twin-lead or coax fittings quickly. Sencore makes a combination FSM that includes a B&W TV screen, and a digital readout.

With a signal strength meter you can measure the TV channel level directly at the antenna terminals, at the output of any amplifier, through splitters and taps, to each outlet. With accurate readings you can plan a new MATV system, or troubleshoot existing ones.

The analog style of FSM has never been easy to use. New technicians have had to spend some time getting familiar with it. The range of the display meter indicator must show you a full-scale reading, ranging from as little as 10 microvolts, to 10,000 microvolts. That is tricky to do. It must also allow you some method of measuring 10,000 μV to perhaps 10,000,000 μV (10 volts), to do this, the meter allows you to insert, via toggle switches, blocks of attenuation, to keep the meter from pegging. A little subtraction or addition are then used to arrive at the resultant signal level.

For example, you have connected the signal wire to the rf coax input on the FSM, and set the band switch to vhf. You start to turn the manual channel selection knob. The meter pegs to the right-hand stop. You flip in one of the 20 dB attenuator switches, the meter still pegs; you flip in another 20 dB of attenuation, (You now have 40 dB of attenuation inserted), the meter now reads -5 dB. The meter is no longer pegged, and trying to burn out the D'Arsonval meter coil. You add 40 dB to the displayed reading of -5 dB, concluding the actual signal level is $+35$ dB. Now you re-center the vhf tuning knob for maximum or peak reading, and find it improves to -2 dB. Now you know the level is actually $+38$ dB. Nearly everyone has difficulty at first with the FSM. It takes a few times of use, then it becomes your best friend. An FSM can help you estimate installation and repair jobs more accurately. It can allow you to accurately aim stationary antennas. An FSM is a must for properly constructing and checking operation of a MATV system.

OFF AIR ANTENNAS

Off air antennas ordinarily have one or more problems. Considering that many were put up by part-time installers, or do-it-yourselfers, the results are amazingly good. If you were somehow to measure the performance of the many roof-top antennas in use today, against the optimum signal possible at each location, the actual-vs-optimum performance percentage would have to be less the 10 percent.

Think about it:

- Over half of the existing antennas in use today in uhf/vhf signal areas have no uhf elements!
- Over half of the fringe-area antennas need a preamp.
- Less than half of the antennas in multi-directional broadcast areas have a rotor.
- Over half of the do-it-yourself antennas, installed with twin lead transmission wire, have the twin-lead securely (and capacitively) taped to a steel mounting pole.
- An estimated 80 percent of all VCR hookups omit the uhf signals entirely.
- At least 5 percent of all uhf/vhf antennas have the uhf driven element still folded in the shipping position.
- At least 10 percent of all antennas are facing the wrong direction.
- At least 25 percent of all antennas are not the proper type for the area, they don't reject ghosts, or they have too little gain.
- At least 50 percent of all twin-lead installations have a lead wire problem.
- Thousands of antennas are in need of physical replacement. The antenna is broken or has missing elements. The tower is rusting away. The guy wires are at the breaking point. Twin-lead has its insulation rubbed off, as it blows back and forth across the roof, chimney, or gutter.

- Thousands of people have spliced both twin-lead and coax directly, thereby reducing the signal and increasing standing waves.
- A majority of today's TV sets are missing uhf/vhf bandsplitters at the TV set.

These examples do not represent an exhaustive list of the problems you might encounter in antenna work, but they do show many common problems existing today. Since antenna theory is complex, since there has been no public education on the topic, and because antennas are sold by radio sales stores, hardwares, and discount merchandisers that encourage set-owners to do-it-themselves, the predictable results are now fact. The good news is that antenna installers can capitalize on the public need for antenna and MATV service.

TROUBLESHOOTING SERVICE CALLS

When contracted to make a repair service call, the technician first should get all the facts from the homeowner:

- What is wrong?
- What stations did you receive prior to this trouble?
- Have you tried a different TV?
- Was the picture snowy or ghosty on any channel before?
- Did the problem occur at the time of a storm?
- Did the problem occur when a new VCR or other item was hooked up?
- After we find the problem and repair it, do you want any improvements? (Coax, amp, rotor, new poles, etc.)
- Do you want to start getting uhf channels?
- Have you ever considered a rotor or dish?
- Will someone be home?
- Are the wires in the crawl space, attic, basement, through the walls?
- How many sets are hooked up now? Do all have the same trouble?

At the location, check the antenna visually. You can see if the uhf driven elements are unfolded into place. Usually the lead wire is connected here too. You might be able to see a broken lead wire. You will immediately see several problems in need of correction: loose standoffs, taped twin-lead, broken elements, wrong antenna direction. Frequently the low-priced, discount store antennas do not grip their elements into the locks. A breeze will blow one loose, laying it across several others, electrically, and intermittently, shorting them out. In a majority of cases you will find a new antenna is necessary.

If all appears to be fine outside, follow the lead wire to the house entry point. Is it open there? Follow it to the amplifier splitter, or TV. If an amplifier is used, check the 14 volt ac output, if it isn't there, you have found the problem. If the 14 volts is, the amplifier on the pole could be dead, change it after checking the output of the power supply, to ascertain that the signal is weak or missing. Keep an eye peeled for bad antennas, twin-lead or coax problems, or moisture in any outside F-connectors. Remember people like to staple wires. They invariably put a staple through the coax or twin-lead, shorting it out.

At the TV set, make sure the signal is connected properly. Sometimes the uhf/vhf splitter leads are reversed. This produces a weak signal on both bands. If all appears fine, hook up your FSM, and mark down on paper the levels for each channel available in the area. This might not seem important, but when you attempt to remember the levels later, you can't. By marking the levels down, you will be able to write the before and after readings down on the customers bill. Doing this, instead of merely saying you got them a better picture, is more professional, plus you will have a record, in case you return for any reason.

If the signal level you find at the TV set is -40 or -30 dB, where previously the customer had a satisfactory signal (0 dB or greater), you have a completely inoperative system. The antenna wire is open, a splitter is bad, the amplifier is dead, or the antenna itself is defective. Climb the tower or roof and verify the antenna is not broken or shorted. If the signal cable is suspect, twist one end together, if twin-lead, and check impedance at the other end with an ohmmeter. Open it to check for a short. (Use your Ohm scale on the DMM.) If coax, check for shorts with both ends open. Check for continuity by putting a barrel splice and 72-ohm terminator on one end, then measure with your digital multimeter.

Splitters installed between the amplifier and its power supply are troublesome. If a dc block is not used in the legs going to TV sets, the uhf/vhf bandsplitter on those sets will short out the 14 V ac power to the preamp. The amplifier power supply has a dc block in its output side. The safe way is to avoid using a splitter on the preamp side, but often it is the best way, saving many extra feet of wire runs. Take care to insert dc blocks where needed, or to use splitters that have only one dc passive port.

INTERFERENCE

There are many types of interference (Chapter 10). Being able to spot it and deal with it, is part of the expertise the reception specialist needs. Owning a spectrum analyzer can help. Knowing the area and the potential causes helps. Contracting with the customer to reward you for solving the problem is an art. Most customers would like for you to do this for free.

Lightning can do strange things to antenna systems. A direct hit can weld the antenna, melt the wire, destroy the preamplifier, and the TV set. Usually it is a jolt that happens several blocks away on a power line. The resultant spike can snap a transistor in the preamp, or open up the stepdown transformer in the power supply. It can zap just one channel on the TV set, leaving the others unaffected. Substituting a known, good TV, will locate this type of defect.

MATV SYSTEM TROUBLESHOOTING

MATV troubleshooting is usually more logical than antenna repair. Because it has several legs, a large portion of the system can be eliminated right off the bat. Dividing the head end up, or substituting another signal from a VCR, or dot/bar generator, will localize the problem. Problems to be expected are:

- buried wires cut by weekend gardeners
- splitters and connectors with moisture problems

- head end problems—dead amps, outside antenna defects, lightning damage.
- direct signal pickup (forward ghosts) usually caused by strong local stations, and a length of unshielded tuner wire, either inside the TV, or from the wall outlet to the TV.

UPGRADING THE MATV SYSTEM

Part of the troubleshooting job can involve improving the existing system. Since many apartments and cable systems were installed years ago, the hardware could be a problem. Usually the uhf channels cannot be received. Be sure the splitters and taps are 900 MHz types. Make sure the head-end is supplying equal uhf and vhf signal levels. If the system is large, and coax replacement is too expensive, consider converting the uhf signals to unused vhf channels. In a majority of apartment installations, the installing electricians used nothing but 17 dB tap-offs. You might be able to make improvements by substituting 12 or 6 dB taps in the tail-end drops. Some variable taps have never been adjusted, and are attenuating the signal a full 17 or 24 dB! You can make improvements by adjusting the taps. The adjustment will be found under the wallplate. Since the tap ordinarily will not adjust to less than −7 dB, there will be cases where you will want to replace the last tap with a splitter, or a 0 dB barrel-outlet.

Some houses have systems using splitters only. You might find converting to taps can improve reception. Additional line amps can be installed if the run is extremely long, and signal level at the head end is too low. Use your FSM to ascertain the size needed. Do not amplify too much because too-large a signal can cause co-channel, or FM interference. In apartments you will also find problems with renters that make wrong hookups or splices, in an attempt to add more outlets.

QUIZ

1. Field Strength Meters are usually powered by:

   a. (   ) batteries
   b. (   ) 110 V ac
   c. (   ) 14 V ac
   d. (   ) solar cells

2. MATV Field Strength Meters should cover the range of:

   a. (   ) −40 to +80 dB
   b. (   ) −20 to +20 dB
   c. (   ) 0 dB to +30 dB
   d. (   ) −100 dB to +100 dB

3. Which of the following cables should not be in close contact with metal objects?

   a. (   ) coax
   b. (   ) twin-lead
   c. (   ) rotor cable
   d. (   ) antenna ground wire

4. On many uhf/vhf all-channel antennas, the twin-lead, or a balun, is connected to wing nuts that are attached to active elements 8 to 12 inches in length. If the installer fails to unfold the elements from the shipping position, parallel to the antenna boom, which channels will be received poorly?

   a. ( ) vhf only
   b. ( ) uhf only
   c. ( ) both uhf and vhf
   d. ( ) FM frequencies only

5. If the FSM check shows +5 dB signals at the wall outlet for all uhf and vhf channels, but the TV picture is still snowy, what is wrong?

   a. ( ) preamp is bad
   b. ( ) preamp power supply is open
   c. ( ) twin lead is probably taped to the mast
   d. ( ) uhf/vhf bandsplitter leads are reversed

6. A uhf/vhf bansplitter usually is a direct short across the transmission line, so far as dc and low frequency ac are concerned.

   a. ( ) true
   b. ( ) false

7. In a motel MATV system, leg one of the two 24-unit legs is dead. Leg two works fine. What could be the trouble?

   a. ( ) the 2-way splitter is open on one side
   b. ( ) wrong size taps were used
   c. ( ) one of the TV sets in leg one has a shorted tuner or balun
   d. ( ) the trunk line amplifier is too weak

8. Vhf channels are distributed well in an MATV system, but local channel 40 is extremely weak. What could be the cause of the problem?

   a. ( ) the trunk line amplifier is too weak
   b. ( ) the 2-way splitter is open
   c. ( ) vhf only splitters and taps are being used
   d. ( ) coax from the preamplifier to the trunk line amplifier must be changed from RG-59 to RG-6.

9. Signal levels are: channel 4 = +20; channel 6 = +25; channel 40 = −15. Adding, 25 dB uhf/vhf amplifier will:

   a. ( ) improve all three channels
   b. ( ) improve channel 40, but cause co-channel interference on channels 4 and 6

c. ( ) cause overload symptoms on all channels
d. ( ) burn out the preamplifier power supply

10. Some MATV coax wall-outlets contain only a chassis-mount barrel splice. Others are:

a. ( ) wall taps
b. ( ) terminators
c. ( ) splitters
d. ( ) traps

Chapter 20

The SAM Checklist

SAM components, parts, and tools, are not readily available everywhere. In fact, supply houses are often located in only a few major cities in each state. A supply house for ordinary electronic parts will likely not have a complete line of MATV parts. Satellite distributors usually do not have antenna parts. If you run out or have need for a different part than that which you carry, it is no easy task to find one.

Because of the unique nature of the tools and parts for SAM work, it is imperative that everything needed for a job be available. It is difficult enough to do the work involved in installing and maintaining the equipment. To find you have to make a return trip to the shop, perhaps 25 or 50 miles, is heartbreaking, not to mention, costly.

To reduce the problem to a minimum, you will find that making a checklist, introducing it to your crews, and religiously using it, is wise. Surely, you will find some additional items that your list needs, and all of the items on these lists are not always required. If you are not doing antenna or MATV work, you can modify the list. It's the same with TVRO. Hopefully, you will want to be in all three categories, and you will then find the antenna list contains most of the equipment.

Table 20-1. Check List for Antenna Parts Inventory.

Antenna Parts List

1. ( ) Spare antennas
2. ( ) Spare pre-amplifiers—4 kinds
   ( ) a. uhf/vhf, coax out
   ( ) b. separate uhf and vhf, coax out
   ( ) c. separate uhf and vhf, twin-lead out
   ( ) d. uhf/vhf combined, twin-lead out
   ( ) e. spare 25 dB line amplifier
   ( ) f. Winegard or other unique mount style amplifiers
3. ( ) Spare 10-foot poles
4. ( ) Rotors
5. ( ) Chimney, gable, peak mounts
6. ( ) Close-wall brackets, guy wire, turnbuckles, eye-bolts
7. ( ) Coax, RG-59, and RG-6
8. ( ) Twin-lead, rotor wire, ground wire
9. ( ) Lightning arrestors (coax and twin-lead), copper ground rods
10. ( ) 3 types of uhf/vhf bandsplitters, baluns, reverse baluns
11. ( ) 2-way, 3-way, 4-way splitters, dc passive splitters
12. ( ) 1-way, 2-way, 4-way tap-offs, 6, 12, 17 dB
13. ( ) A-B switches
14. ( ) 2-antenna joiners into twin and coax
15. ( ) F-connectors for RG-59 and RG-6
16. ( ) Barrel splices and F-F males
17. ( ) Extension cord: 100 feet
18. ( ) Tower materials if applicable
19. ( ) Spare U-bolts, assorted hardware kit, plumbers strapping
20. ( ) 6-outlet ac wall-outlet adapters
21. ( ) wall plates, plugs, brackets
22. ( ) coax terminators
23. ( ) standoff insulators, house feedthrough bushings
24. ( ) 6-foot piece of 3/4-inch pvc pipe
25. ( ) Lag bolts—assorted sizes, lengths
26. ( ) Grounding adapters for 110 V plugs
27. ( ) Video switch box

Table 20-2. Check List for Antenna Tool Inventory.

Antenna Tools Checklist

1. ( ) ladders (40-foot extension and 6-foot step)
2. ( ) ladder braces
3. ( ) 12-foot tree branch snippers
4. ( ) rain gear, gloves
5. ( ) shovel
6. ( ) sealant
7. ( ) coax staple gun, staples, cable clips
8. ( ) hand tools—complete kit with extra 7/16-inch ratchets
9. ( ) 3/8-inch drill, hammer, wood and masonry bits, 12 - 18 inches long
10. ( ) county road maps
11. ( ) lineman's belt
12. ( ) electrician's belt
13. ( ) crimpers
14. ( ) sockets, hacksaw, blades, level
15. ( ) walkie-talkies
16. ( ) flashlights
17. ( ) soldering irons and solder
18. ( ) wire nuts
19. ( ) PVC tape
20. ( ) WD-40 or other lubricant
21. ( ) pipe wrench
22. ( ) hex key set
23. ( ) carpet cutter
24. ( ) other _____

Table 20-3. Check List for Antenna Test Equipment.

Test Equipment

1. ( ) Field Strength Meter
2. ( ) multimeter
3. ( ) hi-pass filter
4. ( ) FM traps
5. ( ) variable attenuator
6. ( ) 3, 6 and 12 dB attenuators
7. ( ) dc blocks
8. ( ) dot/bar generator
9. ( ) portable TV
10. ( ) portable FM radio
11. ( ) topography maps of county
12. ( ) single-channel or band traps (if applicable to your area)
13. ( ) clip leads
14. ( ) fuses, 1A, 3A, 5A, 7A
15. ( ) spare batteries, penlight, C, D, 9V
16. ( ) other _____

Table 20-4. Check List for MATV Parts and Tools.

MATV Checklist
(additions to antenna list)

1. (   ) 35 dB line amplifier
2. (   ) 8-way splitter
3. (   ) cable ties
4. (   ) security sleeves and tool
5. (   ) trencher (if needed)
6. (   ) 7-channel vhf mixer
7. (   ) single channel amplifiers as needed
8. (   ) uhf-to-vhf converter
9. (   ) vhf-to-uhf converter
10. (   ) vhf-to-vhf converter
11. (   ) head-end combiner
12. (   ) single-channel rejection filters as needed
13. (   ) other_____

Table 20-5. Check List for TVRO Parts and Tools.

TVRO Checklist

Site Surveys

1. (   ) inclinometer
2. (   ) compass
3. (   ) county maps
4. (   ) T.I. prone, area location chart
5. (   ) clipboard and paper/pencil (for sketch of yard and entry)
6. (   ) T.I. Feedhorn-LNA-Receiver setup (if needed)
7. (   ) small 55-inch dish setup (if needed)
8. (   ) program guide order blank
9. (   ) sample copy of program guide
10. (   ) call the customer first

TVRO Equipment

1. (   ) receiver
2. (   ) actuator
3. (   ) remote hand unit
4. (   ) actuator motor
5. (   ) proper actuator, saddle clamp, end bolt, bushing, washers
6. (   ) polarizer, and spare
7. (   ) LNA, LNB, LNF, and spare
8. (   ) dish
9. (   ) feedhorn assembly
10. (   ) feedhorn cover

11. ( ) RG-59, RG-6, RG-213 (if needed)
12. ( ) N-connectors, male and female elbows, right angles
13. ( ) scotch locks
14. ( ) concrete (as needed)
15. ( ) downcoverter (if needed)
16. ( ) proper size pole
17. ( ) customer called?
18. ( ) other_____

TVRO Tools

1. ( ) level
2. ( ) cement-mixer, wheel-barrow
3. ( ) tape measure
4. ( ) fish tape
5. ( ) string
6. ( ) cement tools
7. ( ) pvc pipe
8. ( ) special actuator base screw tool, special mount wrenches
9. ( ) trencher
10. ( ) shovel, hoe, spade
11. ( ) gloves, rain gear, boots
12. ( ) pick, posthole-digger
13. ( ) smaller/larger pole shim sleeves
14. ( ) portable welder

Test Gear for TVRO

1. ( ) squawker or tweaker (signal level meter)
2. ( ) spectrum analyzer (if needed)
3. ( ) spare battery-operated, hard-wired intercom with 200 feet of wire
4. ( ) satellite location chart, program guide
5. ( ) 60 - 80 MHz T.I. filters
6. ( ) T.I. troubleshooter kit
7. ( ) distributor/manufacturer service phone numbers
8. ( ) other_____

Chapter 21

Designing a Large MATV System

Chapter 8 encompassed the basic elements of a small MATV system—the type you expect in today's homes and small apartment houses. Other chapters have dealt with the components commonly used, such as splitters, taps, and amplifiers. Antenna technology, the methods of measuring signal levels, and troubleshooting TV signal problems have also been covered. Putting all of this together into a large MATV system, in a hotel, hospital, or mobile home park, isn't much more difficult than expanding the small system. The same rules apply. The same type of hardware is used. The basic differences are:

- Multiunit complexes can require major trenching. This might need to be accomplished while construction is going on.
- Your trenches must be planned, to be compatible with underground electric, plumbing, gas, and other services.
- You could be working in buildings as they are framed.
- The head-end equipment will very likely process each channel prior to insertion into the trunk line, including added features, such as VCR, satellite dish, or locally-generated messages. More and more apartments include video cameras that allow the apartment resident to view anyone entering the building.
- Because many users are involved, poor quality signals on even one channel are not acceptable. The antenna system must clean up all the available channels, and supply higher quality signals than a single family might require.
- The head-end should be accessible, and adaptable to accepting other video/audio sources in the future, such as videotext, and information sources.

MATV SYSTEM ELEMENTS

The following is a list of the various portions of the overall MATV system.

1. The Signal Sources:

   - off-air channels
   - satellite channels
   - local-access channels
   - security channels
   - radio signals

2. The head-end amplifiers and signal processing equipment.
3. The trunk lines and tap-offs to each residence.

The Trunk Line

The trunk line is the source of all the signals into each section, wing, floor, or street in the system. With four separate apartment buildings there will be four trunk lines, all coming off a 4-way splitter. The splitter itself can be supplied by a master trunk line, fed by the head-end equipment.

Because each leg of the system is supplied by a trunk line, care is taken to protect the trunk line from all hazards between the head-end, the building entry, and its path through crawl spaces, attics, and walls. Tap-offs are used to supply each room outlet. The directional tap-offs provide isolation protection against interference generated in a residence. For instance, a resident could hook up a hash-generating computer to the line, or even short his outlet plug. Protection is needed for the trunk line, so that this problem doesn't cause the entire line to go down, or allow interference to be channeled to the other apartments.

If the complex has four duplicate structures, with four apartments each, the system designer has only to calculate the needs of one, and connect all four to a 4-way splitter, after calculating the signal level needed for one. MATV system design can be simplified, usually by dividing it into its separate legs, and calculating the requirements for the longest or highest-loss branch. The other legs will need less signal than the longest, so they will end up with a more ample signal. In the rare case, where the signal required for the longest leg is much too strong for any other leg, simple attenuators can be inserted to optimize the levels.

A large complex of 200 units requires a sophisticated head-end, and signal distribution system. A concept of MATV systems is not immediately recognized: If you have a 200-unit system with a satisfactory signal level, you can add another similar unit, with 200 more units, merely by increasing the signal 3.5 dB, and putting the original trunk line signal into a 2-way splitter!

In systems offering other additional, optional services, at additional monthly cost, the distribution system must be different. It might require a separate coax from the head-end to each apartment. Obviously a large complex will require a massive cable. Two hundred coaxial wires is a large diameter bundle. The head-end junction boxes and switching are also elaborate. In a 16-apartment dwelling, the idea of home run connections is practical. It becomes more impractical as hundreds of units are connected. With the age of addressable decoders upon us, the practical answer to optional programming choices, and even 2-way communications can be available, using the normal trunk line system method.

ONE-LEG LOSSES

The losses in the legs of an MATV system must be calculated. They include:

- splitter losses
- tap-off losses
- wire losses
- tap-off feedthrough, or insertion losses

Splitters

In Chapter 6, you found that splitters are always combinations of 2-way splitters. Two-way splitters have a loss of 3.5 dB per port. Three-way splitters have one port with 3.5 dB and two with 7 dB. Four-way splitters have four outlet ports each, with 7 dB loss. Eight-way splitters have 10.5 dB loss. The losses in a distribution leg are cumulative. If you have four 2-way splitters involved in a leg, you will have four times 3.5 dB, or 14 dB, total splitter loss. Figure 21-1 shows a sample loss calculation for a combination of splitters.

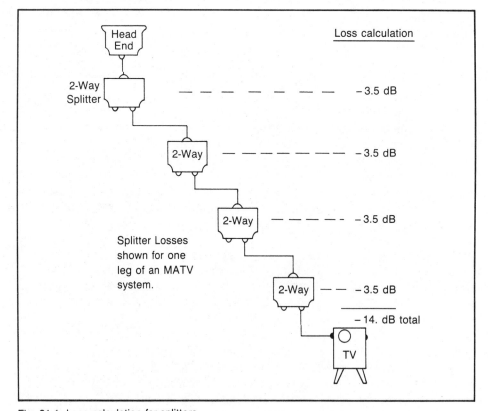

Fig. 21-1. Loss calculation for splitters.

Tap-Off Losses

A tap-off is chosen for each outlet that will reduce the trunk line signal level to an optimum level at the TV outlets in each room. The trunk line can carry a 30 or 40 dB signal level. This amount would overload many TV sets, causing rolling, blackout, or smear. A 24 dB loss tap-off would cut that 30 or 40 dB signal down to +6 to +16 dB, which is optimum. By the same token, if the trunk line has already served a number of tap-offs, and has been reduced in level to +15 to +25 dB, a 24 dB tap might leave too little signal. A smaller value tap should be used. A 12 dB tap would reduce the +15 to +25 dB signal to +3 to +13 dB again an optimum signal level. A common problem is the use of 17 dB taps only. In a small system, with perhaps six line drops, the insertion, or feedthrough loss of the taps will be only 2 or 3 dB total; therefore, the trunk signal level feeding each tap is near the same level, perhaps +24 dB. Seventeen dB taps would be proper for each drop, assuming minimum wire loss was also the case. Each outlet would receive between 0 and +7 dB, a proper signal. If the taps were some distance apart (let's say 100 feet), then the insertion losses combine with the now significant wire losses. After the first tap, the second tap receives only about 18 dB of signal. Using a 17 dB tap, the result at the tap output port is now +1 dB. Taps on down the line each see 6 dB less signal, due to the wire length losses. The third tap on down the line has only 12 dB available. A 17 dB tap leaves a −5 dB signal at the TV. If a 12 dB tap were used, instead of a 17 dB, 0 dB would be available, which is sufficient. The next tap would need to be a 6 dB (tap-off #4). Tap #5 has 0 dB supplied by the trunk now. It would have to be supplied directly to the TV set with no tap loss in order to be sufficient. This isn't a preferred practice. The problem should be solved by increasing the input signal level to the trunk line from 24 dB to 35 dB or more. In that case the tap sizes would be as shown in Fig. 21-2. Note the signal levels are all between 2.4 and +5 dB.

LINE LOSSES

Line losses are important, as you can see. Obviously if you have an extremely long run, line loss can be so large as to require additional line amplifiers to rebuild the signal. Another alternative is to convert uhf signals to vhf channels. This reduces the line loss from a nominal 6 dB per 100 feet, to 4 dB. Another idea is to use amplifiers that have higher gain at higher frequencies. These are called slope amplifiers. Most trunk line amplifiers have gain controls for the three bands (vhf lo, vhf-hi, and uhf) to aid you in optimizing a system.

Calculating the system in Fig. 21-2, but using only vhf channels, we have Fig. 21-3.

HEAD-END DESIGN

With the legs of a system calculated, the head-end can be designed.

If your system has a worst-case leg, having 35 dB loss at the highest channel, and the available signals come off the antenna at a 35 dB level, then you need only the antenna. No amplifier or preamplifier is required. Hook the antenna into the head-end splitter, or trunk line, and the job is completed, other than debugging.

Few areas have 35 dB signals available. More often, a tower is required to get above the terrain, where sufficient signal is available to process. Signal levels below 30 dB on uhf and 40 dB on vhf are so low that the local hash, system, and amplifier-generated

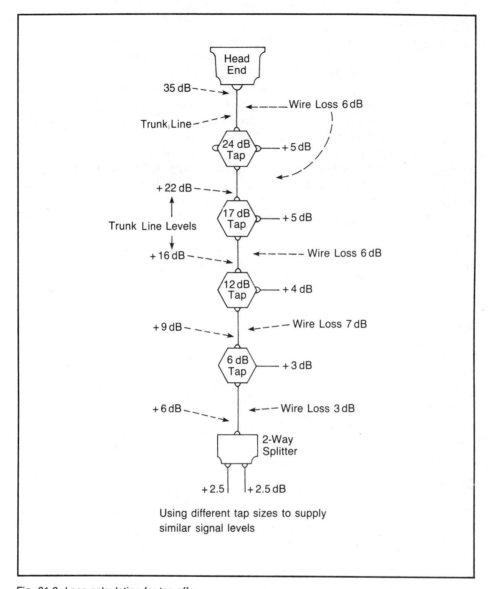

Fig. 21-2. Loss calculation for tap-offs.

noise, are so high that a signal cannot be cleaned up with more amplification. The snow is merely amplified. The solution to the problem is to substitute a dish, use a VCR, or raise the antenna height.

There is a rule of thumb about raising the height of an antenna. It says that for every 10 feet of additional height, the signal should double. Where trees and a diffused signal are all that is available, this rule seems less true. At any rate, if you had a −40 dB signal originally, raising the height 10 feet is supposed to double it to −34 dB. That is only

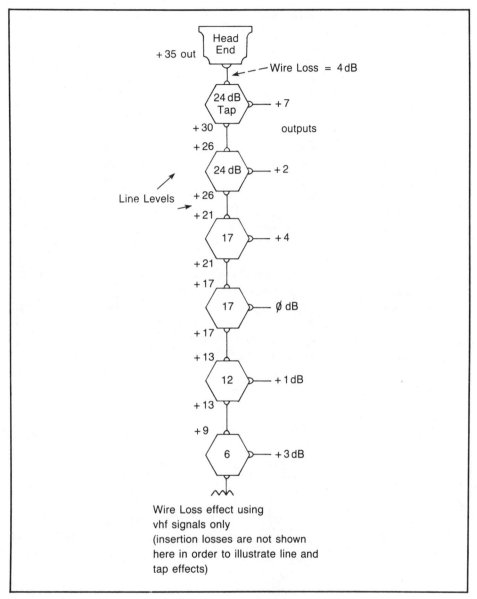

Fig. 21-3. Wire loss calculations in MATV system.

20 microvolts. That is too little. Another 10 feet makes it 40 microvolts, or −28 dB. So if your original antenna was at 30 feet, supplying −40 dB, a 60 foot tower might increase it to −22 dB, or 80 microvolts. (Note all the trouble you went to, to gain 60 microvolts.)

With 80 microvolts, a 17 dB preamplifier will give a −5 dB signal. Feed this into a 35 dB line amplifier and you have a 30 dB signal to supply to the trunks. That still leaves a −5 dB signal at the end of the worst-case line, the one with a 35 dB loss. Solutions are:

- Use a 48 dB distribution amplifier, instead of a 35 dB.
- If you were using a 6 dB tap-off at the last outlet, terminate it with a TV set.
- Add a 6 dB line amplifier somewhere in the worst-case leg.

EQUALIZATION

In MATV systems, the need to supply more channels is common. Since few cities have a dozen channels (all of equal power and direction), the MATV designer must perform a small miracle, providing some channels from a long distance, while utilizing the strong local channels too. This presents a problem of widely varying signal levels. If local channel 2 is strong—say 20 dB, and another channel—#3—is from afar, coming in at −30 dB, you have a problem. Since bandpass filters do not have the ability to chop off frequencies entirely, the edge of the strong channel 2 passband, will interfere with channel 3. If channel 3 were also 20 dB in strength the interference might be negligible. With channel 3 at −30 dB, the edge of channel 2 (the sound frequencies) will be higher than −30 dB, and will swamp out the tiny signal from channel 3.

The solution to such problems is to process the various signals. This can mean reducing a strong one, or amplifying a weak one, converting to a different unused channel, or trapping out the strong channel, until the weak one is built up enough to compete.

Figure 21-4 is an example of signals coming from two different directions, one on the vhf-lo band and one on vhf-hi, both of equal amplitude, both with strong signals levels. This optimum situation is rare. All that is needed, is to combine the signals and supply them to the system.

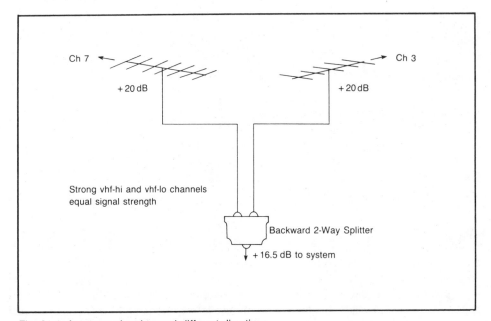

Fig. 21-4. Antennas aimed toward different directions.

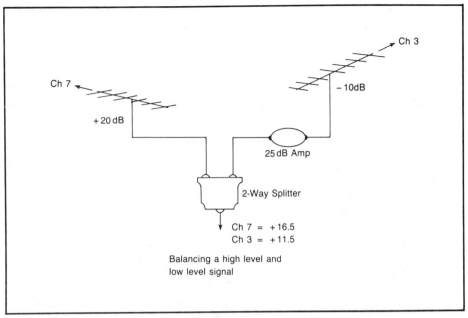

Fig. 21-5. Balancing a 2-direction antenna system.

Figure 21-5 is more typical. One strong channel (7) from one direction, and a weak channel (3) from a different direction. The possibility of ghost reflection pickup of channel 7 on the channel 3 antenna is strong. Using a separate tuned antenna will minimize that possibility. By inserting a 25 dB amplifier in the channel 3 downlead, the signals are brought to a reasonably equal level, suitable for insertion into the trunk line.

The problems mount when you attempt to receive a strong adjacent channel (7) as shown in Fig. 21-6. Channel 8 is weak. Since the two channels are from different

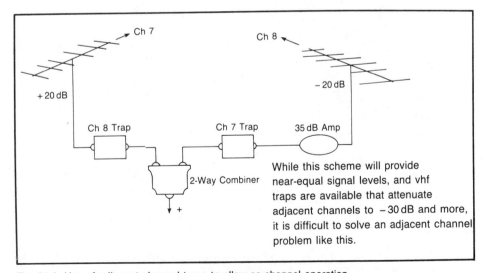

Fig. 21-6. Use of adjacent channel traps to allow co-channel operation.

directions, it is possible to separate the two by using single-channel Yagi antennas, properly oriented. Because the antennas can't be totally frequency-selective, the channel 8 Yagi will be susceptible to picking up some of the channel 7 signal. It will be especially vulnerable to picking up channel 7 reflected signals, since they will be coming from a direction closet to the front of the antenna. In addition to the need to amplify the channel 8 signal, a trap will be needed to reduce the channel 7 high-end (or in this case, the sound) signal that can lap over, and cause adjacent channel interference on 8. The channel 7 sound trap would be used to chop off, or notch out, the channel 7 signal. Success in separating and equalizing the signals will be dependent on the relative directions of channels 7 and 8.

In Fig. 21-7 the problems worsen. Channel 12 has an adjacent channel on both sides. All three channels are of equal amplitude. The best way to reduce adjacent channel interference is to convert channel 12 to another frequency, either vhf or uhf. If the system is able to handle uhf signals, converting 12 to 14 would eliminate adjacent channel interference.

If you have channels 2, 3, or 4 locally, you could experience bleedover, or adjacent channel interference, from products that are output on channel 3 or 4, such as satellite receivers, video disk players, or VCRs. Converting the co-channel, or the satellite output

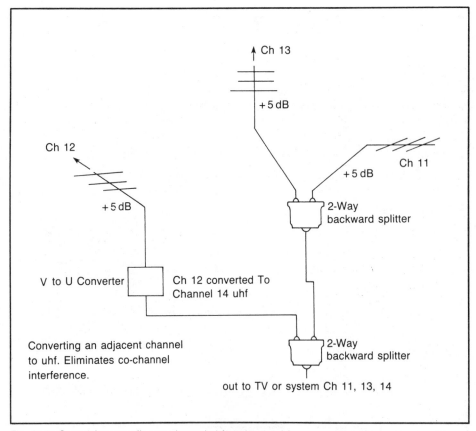

Fig. 21-7. Converting an adjacent channel vhf station to uhf.

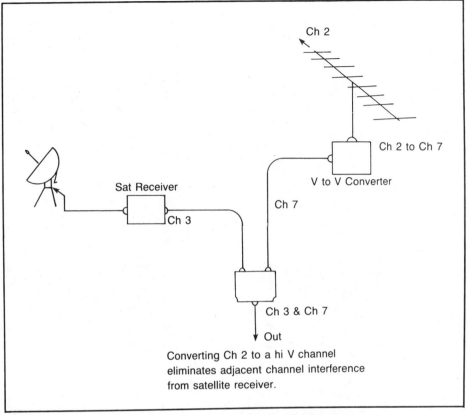

Ch 2

Ch 2 to Ch 7

V to V Converter

Sat Receiver

Ch 3

Ch 7

Ch 3 & Ch 7

Out

Converting Ch 2 to a hi V channel
eliminates adjacent channel interference
from satellite receiver.

Fig. 21-8. Converting a channel to avoid channel 3 satellite interference.

to another unused vhf or uhf frequency would solve the problem. Figure 21-8 illustrates this approach.

Components are made to allow you to solve most problems. Many of the companies listed here will help you design system, or overcome problems. Because the configuration of each system is different, and few localities have identical combinations of broadcast stations, each system has different requirements. There is no standard package one can buy and simply plug in. Economics, geography, and local signal sources, require different, and sometimes unusual, solutions. A beautifully designed and laid-out system can require tuning to make it work right. In addition to the suggested companies that you can choose MA and SMATV hardware from, here is a list of some additional components you might need in order to construct the perfect CATV-MATV system:

- prefabricated or custom head-ends
- single channel Yagi antennas
- single channel, vhf, uhf, or combination, pre-amplifiers
- channel converters, vhf-vhf; uhf-vhf; vhf-uhf
- modulators: CATV, audio/video, TV, audio, or video
- distribution amplifiers
- amplified filters, single channel

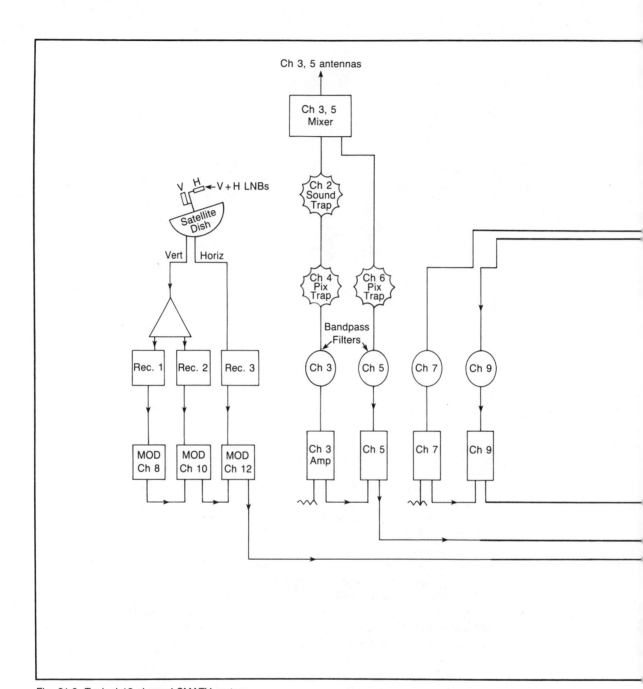

Fig. 21-9. Typical 12 channel SMATV system.

- non-amplified filters, single channel
- passive or active combiners
- band rejection filters
- band joiners

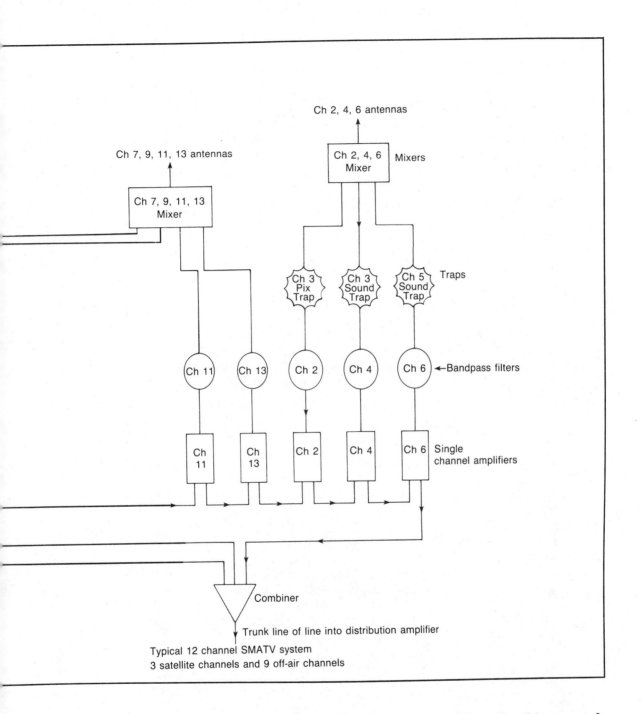

Ch 2, 4, 6 antennas

Ch 7, 9, 11, 13 antennas

Ch 2, 4, 6
Mixer — Mixers

Ch 7, 9, 11, 13
Mixer

Ch 3
Pix
Trap   Ch 3
Sound
Trap   Ch 5
Sound
Trap — Traps

Ch 11   Ch 13   Ch 2   Ch 4   Ch 6 ←Bandpass filters

Ch 11   Ch 13   Ch 2   Ch 4   Ch 6 — Single channel amplifiers

Combiner

Trunk line of line into distribution amplifier

Typical 12 channel SMATV system
3 satellite channels and 9 off-air channels

Figure 21-9 is a typical 12 channel MATV system, with satellite dish support. It uses most of the components required to provide adjacent channels. Note the traps that are needed, and try to figure out why traps are not needed for some channels. Figure 21-10 shows an MATV head-end with 11 channels, and FM radio, using six antennas.

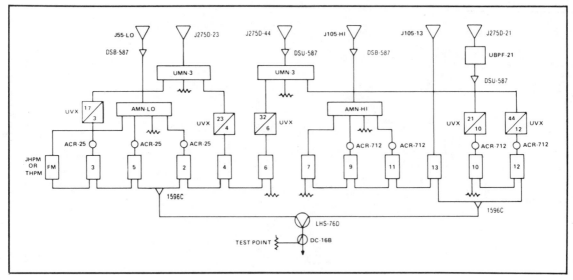

Fig. 21-10. Typical 12 channel MATV system.

QUIZ

1. One major difference between a small MATV system, as might be used for a single-family dwelling, and a large one, serving perhaps a hundred hotel rooms, is:

   a. ( ) the off air antenna(s) are larger on multiple family dwellings
   b. ( ) splitters with more ports are used in large systems
   c. ( ) rotors must be used with larger systems
   d. ( ) wire losses are greater in large systems

2. One reason co-channel, and adjacent channel interference has not been a problem in the past is:

   a. ( ) early TV sets were made to reject adjacent channels and swamp out weak co-channels
   b. ( ) the FCC never assigned adjacent broadcast frequencies in any single area, and kept co-channel assignments at least 200 miles apart.

3. Many MATV systems are designed using separate parallel wiring networks for VCR, closed circuit cameras, and satellite signals.

   a. ( ) true
   b. ( ) false

4. The reasons "homerun" wiring is used in some multiple family dwelling is:

   a. ( ) there is less wire loss in homerun systems
   b. ( ) premium or pay services can be supplied selectively

c. ( ) the extra wire cost is offset by the savings in splitter and taps

d. ( ) 100 outlet tap-offs are now available

5. The number of separate legs in an MATV system is an important factor in determining losses in the longest leg.

a. ( ) true

b. ( ) false

6. An important reason for equalizing signal levels prior to insertion in the trunk line is:

a. ( ) the sound level varies greatly according to the channel signal strength

b. ( ) signals of equal amplitude can reject adjacent channel signals

c. ( ) 10 dB maximum signal deviation is all that modern TV sets can assimilate

d. ( ) splitters will not pass widely-varying signal levels

7. If adjacent channels are to be received, it is usually necessary to:

a. ( ) separate them and trap out adjacent channel frequencies

b. ( ) combine them before amplification

8. One method of eliminating adjacent channel interference is to convert one of the adjacent channels to a different frequency. This is commonly done in large SMATV systems, but rarely in single-family systems. Why?

a. ( ) cost

b. ( ) converters will not work in small systems

9. It is simpler to construct a vhf, or 12-channel MATV system, than a uhf/vhf system, because:

a. ( ) splitter losses are much greater at uhf frequencies

b. ( ) wire losses are much greater at uhf frequencies

10. A radio station or closed-circuit camera could be input to an MATV system by using a:

a. ( ) active filter

b. ( ) signal combiner

c. ( ) mixer

d. ( ) modulator

Chapter 22

The Business Side of SAM

No matter how good you are, your business will not prosper unless you do those things necessary to attract the kind of customers you need. Just about every technician wishes he could be in such demand, that no promotion of the business is required. Those that achieve this condition are indeed rare. "Build a better mousetrap . . ." and people still won't beat a path to your doorstep, unless they know about it.

UNDERCUT YOUR COMPETITION

Undercutting your competition is the most common method of attracting customers. Just advertise a price lower than the competition, and customers will come. That's all you need.

The problem with undercutting is the competition has more experience in the business, and can already buy lower than you, and most likely, he is more efficient. Just because you are a technician, doesn't mean the non-technician installers haven't found shortcuts to lower their installation costs.

About all that undercutting accomplishes, is to attract attention to your company. It also attracts a lower quality of customer, that you will not be able to make a profit from. This can pave the way for your eventual failure.

The best advice is to appear to have low prices, but to maintain a margin on hardware and labor sufficient to pay yourself, and to provide a real profit to the company.

Being known as the highest-priced company is not totally bad; you will attract people who want quality merchandise, and are willing to pay for technical service.

When you undercut, remember it is dangerous. It is better to not do a job, than to do it below cost. It is better to recommend that customers go to your competition, than it is to sell hardware at a break even point, and attempt to do technical service cheap.

WORD OF MOUTH ADVERTISING

Obviously, word of mouth is the cheapest form of advertising. Anything you can do to get recommendations from customers is a plus. Satisfied customers will tell others, just as dissatisfied ones do. For long term business building, do all you can to convince the customer you are the expert. Do all you can to make sure the job is done right technically, and that it looks good. Making a return call to touch up something, or to check out something you thought of later, and weren't convinced was right, at no charge, can surprise a customer, and endear you to them. Picking up your trash, in and out of the house, covering the ditch neatly, or painting the pole, can make a customer yours, and cause him to recommend you to friends and relatives. You don't have to undercut your hardware or labor prices to achieve this.

WAYS TO GET THE BUSINESS

To place a $50 ad in the local newspaper and get immediate response, selling several profitable jobs from it, is everyone's dream. If you can do that, all that is needed for business success, is to do more of it, and increase your work force.

More often, you are faced with a condition where you know you need to spend some money on advertising, but the results frequently produce little or no visible sales. Where some of your efforts pay off, others seem a waste of money. You are tempted to quit advertising completely.

The problem is that satellite, antenna, and MATV systems are not products purchased by everyone, everyday. An antenna is replaced once in ten or so years. A customers' dish will probably be the first he has bought. MATV is often a business-to-business transaction. To expect hoards of customers to race in to your store, for any of these items, is wishful thinking.

My suggestion is to build customer recognition of your firm. By keeping your name in front of the community, you will attract business. In the SAM business, competition is sparse. People quite often have to hunt for someone to repair an antenna, or to do MATV work. You need to give them a way to find you. You need to gently remind them that you are available.

Use a long term approach to advertising. Keep your listing in the classified, and the service directory. Don't fight the number of listings in the yellow pages. Get in all of them pertinent to your business. People look in the yellow-page phone directories to find your kind of business. Rather than blasting the news that you can somehow sell something for $50 less, tell the customer that you are certified, or licensed, or that you can get them more channels, or a clearer picture. Tell them you repair SAM equipment, that you sell rotors, amplifiers and hardware, that you are a reception specialist.

Blow Your Own Horn

You might feel that it is worthwhile to you, and your customer to attend technical SAM schools. It is. That's also a good time to tell the community that you, or your technicians, are going to a seminar. You can get your picture in the paper and some words that mention your firm. It won't get you any immediate customers, but it will help convince them you are a responsible businessman, planning to stay in the SAM business. Some ideas for new releases about your company are:

- the opening of a new store, or change of location
- the addition of a new employee
- certification of one of your technicians
- your election as an officer of your local electronics association
- attendance at a seminar or convention
- an award
- new products
- new services offered

Other Forms of Advertising

High school yearbooks, church bazaars, 4-H, and other local projects might seem like a nuisance, but use them. They build goodwill. If yours is a small town, use radio spots. Advertise service, rather than a product. Billboards can be effective. In most small towns there are bulletin boards at entrances to restaurants and stores. Put your card up. Take a booth at the county fair. You could be the only SAM firm there. Since many people with time on their hands attend fairs, they have time to talk to you. It isn't unusual to acquire two dozen or more jobs from contacts at a fair.

Some dealers feel the service vehicle is not an important advertising tool. It can be an effective reminder to people. Your rolling billboard is in all areas. Your other forms of advertising might not reach everyone. Your neat, clean truck, with its professional signs, can help establish you and your service in the community.

SAM work can be dirty, sweaty, and stressful. The inclination to dress casually is natural. Most customers won't complain if you and your men arrive in jeans and tee shirts. Always keep in mind though, you will get paid more, and with less chance of customer objection, if you dress more professionally than is expected. A good method is to dress well, perhaps in slacks, well shined shoes, clean dress shirt, well groomed, and smelling fresh. Then slip into clean overalls and boots to perform the tough parts of the job. Teach your technicians to do the same. Teach them to respect customer property, and to be courteous.

Hand-Outs

On your way to and from jobs, take the time to stop and leave reminders in newspaper delivery boxes. You will find thousands of broken antennas to leave notes about. If the antenna appears good, leave a flyer, telling about your satellites. Sure some people consider this junk mail. Others appreciate the reminder, and begin thinking about doing something. They will give you the business. Keep flyers that can be used for each different antenna problem in your truck. It's easy to be too busy to do this. The best way is to set a goal of two a day, or ten a day, or one on each trip.

Phone Calls

Most technicians are poor salesmen. Highly trained technical people are rarely good salesmen. You can hire a salesman and solve the problem, or you can understand your natural sales inhibitions, and try to become a good salesperson. One way to do this is simply to do it a lot. After you lose a number of sales and realize certain things you should have said or did, you will improve. Get on the phone, make an established number of phone calls each day. Use a prospect list, or merely get a list out of the phone book. Force yourself to make the calls. After a while you will find most people don't hang up. Some enjoy getting the call, even if they aren't interested in your products. These might later refer you to a friend or relative, just because you called. After a while you will find phone calls are effective in gaining business.

Direct Mail

Always a good form of advertising for small firms, direct mail can be targeted to your previous customers only, or to a prospect list, or to a cold list made up from a county directory. While postage is costly, direct mail is an especially good form of advertising for the SAM dealer. With an inexpensive personal computer, you can make list handling childishly simple, and labels a breeze to print. Since nearly every rural home can use your services, and since SAM products and services are so recognizable, you will find it pays to take the time, and to make the effort, to use direct mail. After you build a list of 200 or more names, you can get a bulk mailing permit at the post office to reduce the postage cost. One new customer might pay you back for all the costs. If you think direct mail is a poor way to advertise, ask yourself why Sears, Radio Shack, and hundreds of insurance companies who use it are so successful.

YOUR EMPLOYEES

The SAM business is not so tough technically that young high school graduates and college students, can't be used for much of the work. These non-technical people often are willing to work for low wages, especially when only part-time. My experience is that a $5 helper is more costly than a $10 technician. There is so much to understand about SAM concepts that the employee should be eager to learn. While there is a place for helpers, a helper who has little interest in SAM work, other than the paycheck, is often costly. While the hourly rate is low, if the helper is never able to work alone, or complete a job properly, you will have to closely oversee his work forever. If he isn't eager to learn, he will never know how to select the proper tap, splitter, use dc blocks, or solve reception problems. If you have to redo his work, or send him back to redo it, the $5 per hour rate ends up really being $15! If his less-than-satisfactory work causes you to lose customers, and your reputation suffers, the cost is even higher. On the other hand, a well-paid technician will try to prove his worth, and will have pride in this work. He will be less likely to dress poorly. Often he is trying to upgrade his knowledge and abilities. All employees must be trained. If you pay the highest wages in town, your employees will stay, rather than continuing the search for a decent job. You will not have to train a new man every three months.

KEEP GOOD RECORDS

It is easy to keep good records in the SAM business. There aren't a great number of transactions, so a monthly profit and loss statement is something you can produce in an hour or less. Get in the habit of making a P and L for each month, and reading it. Make sure you have a proper hardware gross-profit and a labor gross-profit figure, and percentage. Be sure to consider your own time and that of your wife, if she helps in the business. By not including these two items, you will tend to think the business is profitable, when it isn't. This will cause you to underprice hardware, and labor. A hardware gross-profit of 40 percent is not unusual in the MATV antenna business. Satellite sales are a high-ticket item, and often are nearer 30 percent gross profit. Labor gross profit must be 50 percent or more. Overhead costs will vary widely. Depending on taxes and labor costs, overhead can be as low as 25 percent of the gross income.

SOCIALIZE

The SAM business, like electronic repair, and other labor intensive jobs, tends to introvert a person. Once you have become a hermit, contacting other humans only in the course of your daily work, you will establish yourself and your firm as a going-nowhere business. It is easy to quit voting, to resent city, and other government rules. It is easy to classify customers as all trying to beat you. It is easy to resent the policies of suppliers.

To overcome this easy to slip into nitch, join your electronic association. Even if you merely pay your dues, you will start getting valuable information, and you will be asked to comment, to vote, to help on mutual projects. Associations of technicians are good; they encourage mutual help rather than mutual distrust. Association people rarely criticize each other's businesses, while most non-association dealers make a habit of running down their competitors.

Join at least one civic club, the Jaycees, Lions, Rotary, American Legion, etc. The contacts alone will pay the dues. More importantly, broadening of the mind will occur, as you mingle with other people in other businesses. Fellow club members make excellent customers, even though you might think they all expect special favors. Club membership is another way to get local publicity, plus . . . you will have fun.

The SAM business is great, it is more profitable than it has ever been. There is a great demand for it, and few businesses have such a need for qualified workers. A few years ago there was no satellite business. Many television dealers didn't get into it. As the years go by, other opportunities will present themselves to you. Don't close your mind to them. Welcome change, profit from it, keep learning new technology. Keep helping to make the industry better. You can be proud to be a SAM technician or dealer.

Appendix A:

Satellite Dealers, Technician Groups, and SMATV Hardware Suppliers

SATELLITE DEALER AND TECHNICIAN GROUPS

Alabama Satellite Ind. Assn. (AISA)
P.O. Box 4599
Montgomery, AL 36103
(205-834-2001)

Assn. Mexicana De La Industria
De Television Via Satellite
Rio Danubio No. 69 PH
Col. Cuahutemoe, Mexico City 06500

Carolina Dealer Assn.-J. Eddins
Rt. 1 P.O. Box 510
Cheraw, SC 29520

Cent Ohio Sat. Trade Assn.
P.O. Box 18164
Columbus, OH 58727
(614-471-6118)

Electronics Technicians Assn. Int'l.
604 N. Jackson
Greencastle, IN 46135
(317-653-5541)

Florida Assn. of Sat. Dirs.
11791-4 Cleveland Av.
Ft. Myers, FL 33907

Garden State Dlr. Assn.
105 W. 14th St.
Bayonne, NJ 07002
(201-823-1199)

Great Lakes Sat. Dlr. Assn.
3249 Kewaunee Rd.
Green Bay, WI 54301
(414-863-6936)

Independent Sat. Drs. Assn.
304 North Shore Ln.
Culver, IN 46511
(219-842-3105)

Indiana Satellite Drs. Assn. ISDA
4622 E. 10th St.
Indianapolis, IN 46201
(317-357-4575)

Kentucky Sat. Dlr. Assn.
3037 Ring Rd.
Elizabethtown, KY 42701
(502-737-5497)

Louisiana Sat. Dlr. Assn.
P.O. Box 157
Carencro, LA 70520
(318-981-6835)

Mid America Dlr. Assn.
514 A No. 7 Hwy.
Blue Springs, MO 64015
(816-229-1755)

Mississippi Dlr. Assn.
Rt. 9 Box 81A
Philadelphia, MS 39350
(601-656-7501)

New Hampshire Sat. Dlr. Assn.
840 Candia Rd.
Manchester, NH 03103
(603-623-5700)

N. Texas Sat. Dlr. Assn.
2661 Walnut Hill Ln.
Dallas, TX 75229

N. Illinois Sat. Dlr. Assn.
1081 Hwy. 251 S.
Rochelle, IL 61068

NYSAT
114 E. Front St.
Hancock, NY 13783

Ohio Sat. Assn.
33000 Vine St.
Eastlake, OH 44094

Oklahoma Dlr. Assn.
P.O. Box 2315
Edmond, OK 73083

Prince Electronics
1129 Ramey Av.
Worthington, KY 41183

Satellite Comm. Assn. of NW
2918 Portland Rd.
Newberg, OR 97132

Sat. Drs. Assn. of Nevada
6001 W. Spring Mountain Rd. #24
Las Vegas, NV 89102

Sat. Drs. Assn. of Michigan
3125 Twelve Mile Rd.
Berkley, MI 48072

Satellite Ind. Assn.
940 N. Grand Av.
Santa Ana, CA 92701

Sat. Tact Action Ret. Soc.
2540 E. Glendale
Sparks, NV 89431

Sat. TV Drs. Assn. of GA
1447 Peachtree St., NE #804
Atlanta, GA 30309

Sat. TV Systems of NJ & NY
1975 Utica Av.
Brooklyn, NY 11234

So. Utah Drs. Assn.
P.O. Box 12
Beaver, UT 84713

SW Sat Drs Assn.
11135 Dyer St.
El Paso, TX 79934

STRATA
3250 Lincoln Way, E.
Massillon, OH 44646

Tenn/Ark/Miss Dlr. Assn.
3536 Canada Rd.
Lakeland, TN 38002

Tenn. Sat. Dlr. Assn.
3804 Oxford St.
Nashville, TN 37216

Tri-State Sat. Dlr. Assn.
Rt. 73 P.O. Box 174
Berlin, NJ 08009

United Sat. Assn. of Ariz.
8518 N. 7th St.
Phoenix, AZ 85020

United Sat. Assn. of PA
7149 Sterling Rd.
Harrisburg, PA 17112

Western Mass. Dlr. Assn.
396 Main
E. Hampton, MA 01027

Women in Sat. Comm.
1920 North St., NW Suite 510
Washington, DC 20036

SOME SMATV HARDWARE SUPPLIERS

Blonder Tongue
1 Old Jake Rd.
Old Bridge, NJ 08857
(201-679-4000)

Channel Master
P.O. Box 1416
Smithfield, NC 27577
(919-934-9711)

General Instruments
1 Taco Street
Sherburne, NY 13400
(607-674-2211)

Pico-Macom
12500 Foothill Bl.
Lakeview Terrace, CA 91342
(800-421-6511)

Winegard
3000 Kirkwood St.
Burlington, IA 52601
(319-753-0121)

Appendix B:
Satellite Channel Choices

ENTER ACTUATOR NUMBER

TRANSPONDER	SATCOM 2R 72°W [F2]	GALAXY 2 74°W [G2]	SATCOM 4 83°W [F4]	TELSTAR 302 87.5°W [T2]	WESTAR 3 91°W [W3]	GALAXY 3 93.5°W [G3]	TELSTAR 301 96°W [T1]	WESTAR 4 99°W [W4]	ANIK D 104.5°W [AD]	ANIK [109°W [
1		sports feeds	Home Shopping Net 2	news, sports feeds			ABC, CBS feeds		news-sports	
2	occasional feeds	sports feeds	Bravo (s.s.5.8 mpx)			sports feeds Horse Racing (enc.)	CBS Network programs (East)	ABC, CNN, NBC news, sports feeds	The Sports Network	
3		sports feeds	Rock Christian Net				ABC N.Y./Chi		Independent & CBC feeds	
4		sports feeds	Nickelodeon (West)			sports feeds	ABC, CNN sports feeds	sports feeds	Global TV	
5		sports feeds			CNN feed	sports feeds	sports feeds	Bonneville Sat feed	CBC sports feeds	
6			MSG Network TelShop			sports feeds	ABC-TV N.Y Nightline feeds	Wold Bay Meadows (enc.)	MuchMusic (s.s.5.41-6.2 nd)	
7			Liberty Broadcasting			news feeds	CBS Network programs (East)		Winnipeg CBC feeds	sports feed
8		sports feeds	C-SPAN II U.S. Senate	occasional feeds		news, sports feeds	occasional feeds		CHCH Hamilton (enc.)	
9			SportsVision	sports feeds		news, sports feeds	Wold Communications	CTNA (enc.) news, sports feeds	WDIV-NBC Detroit (enc.)	
10		sports feeds	American Movie Classics	ABC Network programs (West)		The Meadows Racing news, sports feeds	ABC Network programs (East)	sports feeds	WXYZ-ABC Detroit (enc.)	
11			Home Sports Ent.	ABC Network feeds		sports feeds	sports feeds	news, sports feeds	CBC North (Pac.)	sports feed
12			The Sports Channel (N.Y.)			Rep. News Conf sports feeds	ABC Network programs (East)	Home racing occasional		
13	NASA feeds	sports feeds	New England Sports Network	occasional feeds		Sky Merchant sports feeds	news, sports feeds	JISO feeds	news & sports feeds	sports feed
14			TV Shoppers Showcase Crazy Eddie	Wold Communications feeds			ABC East	news, sports feeds	TCTV Montreal (enc.)	
15		sports feeds	Shop at Home	CBS News feeds			CBS News feeds sports feeds	PBS-A	CBC-French	sports feed
16		sports feeds		CBS Network programs (West)			CBS feeds	CNN feeds	(CBC) House of Commons French	
17		sports feeds		CBS Network programs (East)		news, sports feeds	occasional feeds	PBS-B	Visnews/London CBC feeds	sports feed
18		sports feeds	Hit Video USA (s.s.5.8-6.8 nd)	CBS feed		Horse Racing (enc.) sports feeds	occasional feeds	Wold sports feeds	GTV Edmonton (enc.)	
19			WPIX N.Y.	CBS News feeds	CBS, NBC news, sports feeds		occasional feeds	Bonneville/ITN feeds CBS sports	CBC North (Atl.)	
20		sports feeds	Prime Ticket Sports CDN 1	CBS Network programs (East)				America's Mkt. Place Horse Racing (enc.)	CBMT Montreal	
21			Nostalgia Channel	Netcom feeds	sports feeds	Horse Racing (enc.)	Netcom feeds	PBS-C	WTVS-PBS Detroit (enc.)	
22	Armed Forces Sat. Net (AFRTS)		Home Team Sports	CBS News sports feeds		occasional feeds	sports feeds	news, sports feeds	CHAN Vancouver B.C (enc.)	
23		sports feeds	New Eng. Spts Ch The Silent Network	CBS feeds	sports feeds		Wold Satellite Net	PBS news, sports feeds	WJBK-CBS Detroit (enc.)	
24			CDN 2 The Playboy Channel	CBS feeds		occasional feeds	TVSC Pittsburgh feeds	ABC, CBS sports feeds	(CBC) House of Commons English	

NOTES:

(enc.) = Encrypted
s/s = Sound Subcarrier
actuator number = Reference Number from Motor Drive
Westar 2 & 3, Anik B are 12-transponder satellites.

mpx = Multiplex   n/d = Narrow Discrete
w/mat = Wide Matrix   w/d = Wide Discrete

*Programming may soon be carried solely on Ku-band (12 GHz) not normally receivable on C-band home satellite earth stations.

	MORELOS 1 113.5°W [M1]	TRANSPONDER	SPACENET 1 120°W [S1]	WESTAR 5 122.5°W [W5]	TELSTAR 303 125°W [T3]	ASC-1 128°W [A1]	SATCOM 3R 131°W [F3]	GALAXY 1 134°W [G1]	SATCOM 1R 139°W [F1]	AURORA SATCOM 5 143°W [F5]	TRANSPONDER
		1	sports feeds	ABC Brightstar	Country Music TV (s.s.5.58/5.76 nd)		Nickelodeon (East)	HBO (East) (enc.)	occasional feeds		1
	XHITM Mexico City	2	sports, CNN feeds	University Net Dr. Gene Scott	SelecTV (enc.)		The Learning Channel Home Shopping Net 2	Nashville Network (s.s.5.58-5.76)			2
		3	American Extasy (enc.) sports/WLVI	occasional feeds		Nite Line TV Net	Trinity Broadcasting	WGN-Chicago			3
	KFMB San Diego	4	sports, news feeds	occasional feeds			Telshop FNN/SCORE	Disney Channel (East) (s.s5.8-6.8)			4
		5	Christian Television Network					Showtime (East) (enc.)	NetCom/Globo NBC News feed		5
		6	occasional feeds				Tempo Network	SIN			6
	sports feeds	7		CBS sports feeds			ESPN (s.s.5.58-5.76 nd)	CNN (enc.)	Prime Ticket Sports FNN		7
	XHDF Mexico City	8	sports feeds	Pro Am Sports Net INN News feeds	news, sports feeds		CBN Cable Network	CNN Headline News (enc.)	NBC Network East (s.s.5.8-6.8 nd)		8
	sports feeds	9	First Run PPV (enc.) Health Net/Boresight				USA (West) (s.s.5.58-5.76 nd)	ESPN (s.s.5.58-5.76 nd)			9
		10					Showtime (West) (enc.)	Movie Channel (enc.) (s.s.5.58/5.76) (East)			10
		11	Hospital Satellite Network (enc.)	occasional feeds			MTV (s.s.5.8-6.62 w/mat)	CBN Cable Network	WTN/London to Australia feeds		11
		12					EWTN		Netcom NBC Sports		12
		13	sports feeds				HBO (West) (enc.)	C-SPAN U.S. House			13
	XEW Mexico City	14		BBC, CBS, INN News feeds			Cable Value Network	Movie Channel (enc.) (s.s.5.8-6.8)(West)			14
	sports feeds	15	ACTS	Horse Racing (enc.) sports feeds			VH1 digital stereo (s.s.7.4)	WORN J (enc.)	Visnews London		15
		16	SIN news sports	Caribbean Superstation			Home Shopping Net 2 HTN Plus (s.s.6.8 mpx)				16
		17	occasional feeds				Lifetime	PTL (s.s.5.58-5.76 nd)	news feeds		17
	sports feeds	18		Group W BBC News sports			Reuters Videotext	WTBS (s.s.5.58-5.76 nd)	NBC feeds		18
		19	Shopping Line	occasional feeds			Weather Channel	Cinemax (East) (enc.)			19
		20	occasional feeds	Telebet Racing (enc.) sports feeds	ABC, NBC, CBS Honolulu feeds		BET (s.s.6.8-7.38 nd)	Galavision		occasional feeds	20
		21	BTN (enc.)	Inday news feeds				USA (East) (s.s.5.58-5.76 nd)	Netcom		21
		22		Taft Broadcasting Scrambling Info. Chan.	KTVT-Dallas		Home Shopping Network 1	Discovery Network	sports feeds		22
	sports feeds	23	news, sports feeds	sports feeds			Cinemax (West) (enc.)	HBO (East) (enc.)			23
		24		sports feeds			Arts & Entertainment (s.s.5.58-5.76 nd)	Disney Channel (West) (s.s.5.8-6.8)	Visnews Radio Italiana	Alaska Satellite TV Project	24

⌐ = Horizontal Polarity   ▦ = Vertical Polarity

satellite TV week

Continental U.S.: 800-345-TVRO (8876) • Others: 707-725-1185
P.O. Box 308 • Fortuna, CA 95540

Appendix C:

The Professional Electronics Technicians Association Membership Application

Membership Application

THE PROFESSIONAL ELECTRONICS TECHNICIANS ASSOCIATION

Name_____ Phone_____
(First)   (Middle)   (Last)

Address_____ City_____ State_____ Zip_____

Age_____

Present Employer_____ Position_____

Years in Position_____

Types of Activity You Are Now Engaged In:

[ ] Industrial
[ ] Education
    ( ) Instructor ( ) Student
[ ] Military
[ ] Medical
[ ] Consumer
[ ] Sound
[ ] Communications/TVRO
[ ] Computers/Office & Equipment
[ ] Data Transmission
[ ] Mining/Chemicals
[ ] CCTV/MATV/Antennas
[ ] Broadcasting
[ ] Engineering
[ ] Musical Instruments
[ ] Distribution
[ ] Other _____

In addition to regular ETA membership, check the division you wish to be a part of:

[ ] Educators (EEA)
[ ] Certified Technicians (CTD)
[ ] Canadian Division (ETA-C)
[ ] Communication Techs (CD)
[ ] Medical (BMD)
[ ] Industrial (ID)
[ ] Shopowners (SO)

Enjoy these important benefits as a member of ETA

Membership Decals
ETA Member Wall Certificate
Wallet Identification Card
By Laws
Monthly Association News
Monthly Tech Training Program
Monthly Management Update
Area Technical & Business seminars
Discounts on Tech Publications
Job Placement Assistance
Certification Examinations
Small Business Administration Assistance
PR Brochures
Life & Health Insurance Savings
Annual Technician Convention
Help when you need it.

ETA ANNUAL DUES

Institutions . . . . $125.00 per year
Business Owners 30.00
Employee Technician 25.00
Electronics Student 15.00

Name of School You Are Enrolled In:

_____

Please Sign This Application and return with your fee:

Signed _____

Date_____ Amt. $_____

OTHER INFORMATION:

Please give your CET No. or FCC No. if

Applicable: _____ _____
(Not Required)

Other Registrations or Honors: _____

_____

_____

Mail to . . .

ELECTRONICS TECHNICIANS ASSOCIATION
604 NORTH JACKSON STREET
GREENCASTLE, INDIANA 46135
(317) 653-3849

Appendix D: ETA Information

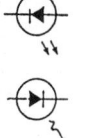

JOURNEYMAN: for technicians with four or more years of combined schooling and/or experience. The exam consists of taking the ASSOCIATE portion, PLUS an option of your choice—the two sections making up one complete journeyman exam. (A technician registered as an associate CET takes only the journeyman option.) This exam also requires a 75% passage rate. Upon passage you will receive the CERTIFIED ELECTRONICS TECHNICIAN certificate and wallet card.

SENIOR: Those technicians that have at least eight years in the profession, can take the journeyman test, and with a score of at least 85% can receive the SENIOR CERTIFIED ELECTRONICS-TECHNICIAN certificate and wallet card. A brass plaque is also awarded for this level.

MASTER: Technicians who are unique in being proficient in all six major technical fields can take all six options at once, and if successful, and having eight or more years experience, may receive a MASTER CERTIFIED ELECTRONICS TECHNICIAN and wallet card. A brass plaque is also awarded for this level. A score of 75% is required.

What are the options?

With the many areas of specialty that electronics technicians can work in, defining categories is somewhat of a problem. However, the certification exam is currently divided into six (6) major categories. Three of the categories are further divided into specialties. The categories and specialties are:

1. Consumer
   —Radio-Television
   —Audio Hi-Fi

2. Commercial
   —MATV
   —VCR (VTR)
   —TVRO (Satellite)

3. Communications
   —Two-Way
   —Avionics

4. Industrial

5. Computer

6. Bio-Medical

Take the option-specialty that best suits your work.

Test Fees (effective August 1, 1986)

Associate	$20	Senior	$40
Journeyman	$30	Master	$65

☐ Yes, I want to take the ETA certification exam. Please send the name and address of the closest certification administrator.

☐ Please send me TAB BOOK'S THE CET EXAM BOOK #1670. A study guide by Dick Glass and Ron Crow. Enclosed is $11.95 (includes shipping).

☐ I want to join the Electronics Technicians Association. Please send an application.

ETAI Testing
P.O. Box 1258
ISU Station
Ames, Iowa 50010

Your Name _____

Address _____

City, State, Zip _____

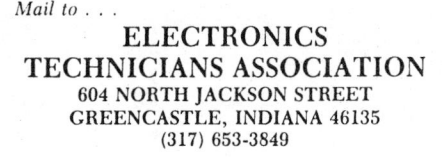

Mail to . . .
ELECTRONICS
TECHNICIANS ASSOCIATION
604 NORTH JACKSON STREET
GREENCASTLE, INDIANA 46135
(317) 653-3849

PROFESSIONALS

. . . Are Certified

by

THE ELECTRONICS TECHNICIANS
ASSOCIATION, INTERNATIONAL

Who should become certified?

Every technician who plans to make electronics a career should take the exam and be registered with ETA as a CET. Not to do so is to continue to the status quo that has placed electronics technicians in low income categories and left many in low-respect jobs. Certification won't turn that around overnight, but each time a technician does become certified, the odds turn a little bit more in your favor.

Who can become certified?

Any electronics student or technician. You don't have to have a trade school degree, and you don't have to join ETA or any other group.

When do I take the exam?

ETA has monitors in most locales. Where we don't have one we can arrange for testing at a local electronics educational institution.

Who makes up the ETA exam?

From our certification headquarters, Ron Crow of Iowa State University in Ames, and his staff assemble examinations based on modern electronics technology theory and practices submitted by active technicians who work as an exam committee. Because ETA (a non-profit association) is organized and directed by independent technicians, the exam is most practical.

If I'm an efficient technician, why do I need certification?

It would be extremely rare to find a technician who can't pass the CET exam, but who is a crackerjack technician. Over the years, the exam has done a beautiful job of separating the "cut and try" technicians from really good troubleshooters.

How can certification help me make more money?

The first way is in bringing more respect to the profession as a whole. Your recognition and eventual income will gain as more awareness of the program is gained. Without certification you are a loner, most often receiving what others want to give you.

What can I study to help prepare for the exam?

You can become a member of ETA and receive monthly training programs that help, and you can order all back issues of these monologues for a small fee. You can get a group of your fellow technicians together and rent ETA video tapes that are excellent study aids.

Also available is a study guide:
The CET Exam Book
Dick Glass and Ron Crow
Book #1670 TAB Publishers
This book should be available at your local electronics parts outlet or order from ETA using the order form on the last page of this brochure.

What can teachers and students do?

Electronics instructors can encourage technicians to take the exam. They can help the students set up study sessions prior to taking the associate exam. The test can be given at the school as long as an outside ETA monitor assists. You may also contact Dick Glass, SrCET, President of ETA, 604 N. Jackson, Greencastle, Indiana 46135 (317/653-3849) for information on forming a student chapter.

I'm in the military. How do I take the exam?

It is the easiest way of all. DANTES—The Defense Activity for Non-Traditional Education Support—encourages certification by offering the ETA exam in every military installation in the world. Just have your DANTES test officer request an exam from ETA on his/her official stationery.

I've been certified for years. Is my registration still valid?

Yes. If you could pass the CET exam previously, you are still recognized as a CET. Many technicians, realizing that technology has changed, want to update their certification. ETA has an answer. The Senior CET program allows technicians with 8 or more years experience to retake the updated version of the exam. If successful (85% is a passing grade), you are then certified and registered with ETA as a SENIOR CET.

Four levels of exams

ASSOCIATE: for students and technicians with less than four years of experience. The exam is on basic electronics technology and requires a score of 75% to pass. Upon passage you will receive the ASSOCIATE ELECTRONICS TECHNICIAN wallet certificate and wallet card. (Registration expires after four years.)

Appendix E: TV Frequencies

Channel Number	Frequency Limits of Channel	Center Frequency of Carrier		CATV

SUB CHANNEL

Channel	Freq. Limits	Picture	Sound	CATV
A	18MHz – 24MHz	19.0	23.5	T-9
C	30MHz – 36MHz	31.0	35.5	T-11
E	42MHz – 48MHz	43.0	47.5	T-13

VHF LOW BAND

Channel	Freq. Limits	Picture	Sound
2	54MHz – 60MHz	55.25	59.75
3	60MHz – 66MHz	61.25	65.75
4	66MHz – 72MHz	67.25	71.75
5	76MHz – 82MHz	77.25	81.75
6	82MHz – 88MHz	83.25	87.75

FM BAND

Freq. Limits	Center Freq.
88MHz	88.00
108MHz	108.00

MID BAND

Channel	Freq. Limits	Picture	Sound
A	120MHz – 126MHz	121.25	125.75
B	126MHz – 132MHz	127.25	131.75
C	132MHz – 138MHz	133.25	137.75
D	138MHz – 144MHz	139.25	143.75
E	144MHz – 150MHz	145.25	149.75
F	150MHz – 156MHz	151.25	155.75
G	156MHz – 162MHz	157.25	161.75
H	162MHz – 168MHz	163.25	167.75
I	168MHz – 174MHz	169.25	173.75

VHF HIGH BAND

Channel	Freq. Limits	Picture	Sound
7	174MHz – 180MHz	175.25	179.75
8	180MHz – 186MHz	181.25	185.75
9	186MHz – 192MHz	187.25	191.75
10	192MHz – 198MHz	193.25	197.75
11	198MHz – 204MHz	199.25	203.75
12	204MHz – 210MHz	205.25	209.75
13	210MHz – 216MHz	211.25	215.75

SUPER BAND

Channel	Freq. Limits	Picture	Sound
J	216MHz – 222MHz	217.25	221.75
K	222MHz – 228MHz	223.25	227.75
L	228MHz – 234MHz	229.25	233.75
M	234MHz – 240MHz	235.25	239.75
N	240MHz – 246MHz	241.25	245.75
O	246MHz – 252MHz	247.25	251.75
P	252MHz – 258MHz	253.25	257.75
Q	258MHz – 264MHz	259.25	263.75
R	264MHz – 270MHz	265.25	269.75
S	270MHz – 276MHz	271.25	275.75

SUPER BAND (Cont.)

Channel	Freq. Limits	Picture	Sound
T	276MHz – 282MHz	277.25	281.75
U	282MHz – 288MHz	283.25	287.75
V	288MHz – 294MHz	289.25	293.75
W	294MHz – 300MHz	295.25	299.75

UHF BAND

Channel	Freq. Limits	Picture	Sound
14	470MHz – 476MHz	471.25	475.75
15	476MHz – 482MHz	477.25	481.75
16	482MHz – 488MHz	483.25	487.75
17	488MHz – 494MHz	489.25	493.75
18	494MHz – 500MHz	495.25	499.75
19	500MHz – 506MHz	501.25	505.75
20	506MHz – 512MHz	507.25	511.75
21	512MHz – 518MHz	513.25	517.75
22	518MHz – 524MHz	519.25	523.75
23	524MHz – 530MHz	525.25	529.75
24	530MHz – 536MHz	531.25	535.75
25	536MHz – 542MHz	537.25	541.75
26	542MHz – 548MHz	543.25	547.75
27	548MHz – 554MHz	549.25	553.75
28	554MHz – 560MHz	555.25	559.75
29	560MHz – 566MHz	561.25	565.75
30	566MHz – 572MHz	567.25	571.75
31	572MHz – 578MHz	573.25	577.75
32	578MHz – 584MHz	579.25	583.75
33	584MHz – 590MHz	585.25	589.75
34	590MHz – 596MHz	591.25	595.75
35	596MHz – 602MHz	597.25	601.75
36	602MHz – 608MHz	603.25	607.75
37	608MHz – 614MHz	609.25	613.75
38	614MHz – 620MHz	615.25	619.75
39	620MHz – 626MHz	621.25	625.75
40	626MHz – 632MHz	627.25	631.75
41	632MHz – 638MHz	633.25	637.75
42	638MHz – 644MHz	639.25	643.75
43	644MHz – 650MHz	645.25	649.75
44	650MHz – 656MHz	651.25	655.75
45	656MHz – 662MHz	657.25	661.75
46	662MHz – 668MHz	663.25	667.75
47	668MHz – 674MHz	669.25	673.75
48	674MHz – 680MHz	675.25	679.75
49	680MHz – 686MHz	681.25	685.75
50	686MHz – 692MHz	687.25	691.75
51	692MHz – 698MHz	693.25	697.75
52	698MHz – 704MHz	699.25	703.75
53	704MHz – 710MHz	705.25	709.75
54	710MHz – 716MHz	711.25	715.75
55	716MHz – 722MHz	717.25	721.75

UHF BAND (Cont.)

Channel	Freq. Limits	Picture	Sound
56	722MHz – 728MHz	723.25	727.75
57	728MHz – 734MHz	729.25	733.75
58	734MHz – 740MHz	735.25	739.75
59	740MHz – 746MHz	741.25	745.75
60	746MHz – 752MHz	747.25	751.75
61	752MHz – 758MHz	753.25	757.75
62	758MHz – 764MHz	759.25	763.75
63	764MHz – 770MHz	765.25	769.75
64	770MHz – 776MHz	771.25	775.75
65	776MHz – 782MHz	777.25	781.75
66	782MHz – 788MHz	783.25	787.75
67	788MHz – 794MHz	789.25	793.75
68	794MHz – 800MHz	795.25	799.75
69	800MHz – 806MHz	801.25	805.75

TRANSLATOR FREQ.

Channel	Freq. Limits	Picture	Sound
70	806MHz – 812MHz	807.25	811.75
71	812MHz – 818MHz	813.25	817.75
72	818MHz – 824MHz	819.25	823.75
73	824MHz – 830MHz	825.25	829.75
74	830MHz – 836MHz	831.25	835.75
75	836MHz – 842MHz	837.25	841.75
76	842MHz – 848MHz	843.25	847.75
77	848MHz – 854MHz	849.25	853.75
78	854MHz – 860MHz	855.25	859.75
79	860MHz – 866MHz	861.25	865.75
80	866MHz – 872MHz	867.25	871.75
81	872MHz – 878MHz	873.25	877.75
82	878MHz – 884MHz	879.25	883.75
83	884MHz – 890MHz	885.25	889.75

Answer Key

Chapter 1

1. a
2. a
3. a
4. a
5. b
6. b
7. b
8. b
9. e
10. a

Chapter 2

1. b
2. b
3. a
4. a
5. a
6. b
7. a

8. a
9. b
10. b

Chapter 3

1. b
2. a
3. a
4. a
5. a
6. a
7. a
8. a
9. c
10. b

Chapter 4

1. a
2. b
3. d

4. b
5. b

Chapter 5

1. b
2. d
3. b
4. a
5. a

Chapter 6

1. a
2. a
3. d
4. b
5. a
6. d
7. b
8. b
9. d
10. a

Chapter 7

1. d
2. d
3. b
4. c
5. b
6. b
7. c
8. b
9. c

Chapter 8

1. a
2. b
3. a
4. d
5. d
6. a
7. b
8. b
9. b
10. c

Chapter 9

1. b
2. a
3. b
4. b
5. b
6. b
7. b
8. b
9. c
10. d

Chapter 10

1. a
2. b
3. b
4. c
5. b
6. b

7. c
8. b
9. b
10. b

Chapter 11

1. b
2. c
3. a
4. c
5. d
6. b
7. b
8. b
9. b
10. a

Chapter 12

1. a
2. b
3. b
4. a
5. a
6. b
7. c
8. b
9. a
10. a

Chapter 13

1. a
2. d
3. a
4. b
5. a
6. b
7. a
8. b
9. a
10. b

Chapter 14

1. b

2. b
3. a
4. b
5. a
6. d
7. b
8. b
9. a
10. b

Chapter 15

1. b
2. a
3. c
4. b
5. d
6. a
7. b
8. b
9. a
10. a

Chapter 16

1. c
2. b
3. a
4. a
5. a
6. b
7. b
8. b
9. b
10. b

Chapter 17

1. a
2. a
3. c
4. a
5. a
6. d
7. a
8. c
9. a
10. a

Chapter 18	Chapter 19	Chapter 21
1. a	1. a	1. d
2. b	2. a	2. b
3. d	3. b	3. b
4. c	4. b	4. b
5. b	5. d	5. b
6. a	6. a	6. b
7. b	7. a	7. a
8. c	8. c	8. a
9. d	9. b	9. b
10. d	10. a	10. d

Index